TELEPEN

AF616235

APPLIED SURFACE MODELLING

APPLIED SURFACE MODELLING

C. F. M. CREASY B.Sc., M.Sc., Ph.D.
Principal Lecturer
Department of Mathematics, Newcastle upon Tyne Polytechnic

C. CRAGGS B.Sc., Ph.D.
Principal Lecturer
Department of Mathematics, Newcastle upon Tyne Polytechnic

ELLIS HORWOOD
NEW YORK LONDON TORONTO SYDNEY TOKYO SINGAPORE

First published in 1990 by
ELLIS HORWOOD LIMITED
Market Cross House, Cooper Street,
Chichester, West Sussex, PO19 1EB, England

A division of
Simon & Schuster International Group
A Paramount Communications Company

Typeset in Times by Ellis Horwood Limited
Printed and bound in Great Britain
by Hartnolls, Bodmin, Cornwall

British Library Cataloguing in Publication Data

Creasy, C. F. M.
Applied surface modelling.
1. Surfaces. Simulations models.
Applications of computer systems.
I. Title. II. Craggs, C.
516.36011
ISBN 0–13–040007–6

Library of Congress Cataloging-in-Publication Data

Creasy, C. F. M. (Colin F. M.), 1950–
Applied surface modelling / C. F. M. Creasy, C. Craggs.
p. cm.
ISBN 0–13–040007–6
1. Geometry — Data processing — Congresses. 2. Surfaces — Mathematical models — Congresses. 3. Computer-aided design — Congresses. I. Craggs, C. (Carolyn), 1954– . II. Title.
QA448.D38C74 1990
516′.001′13–dc20 90–4505
CIP

Table of contents

Preface

The chapters of this volume were originally papers presented orally at the twelfth POLYMODEL conference entitled 'Applied surface modelling' held at Newcastle upon Tyne Polytechnic in May 1989. The conferences are organized annually by the North East of England Polytechnic's Mathematical Modelling and Computer Simulation Group — POLYMODEL. The Group is a non-profitmaking organization based on the mathematics departments of the three polytechnics in the region, and has a membership drawn from those educational institutions and from regional industry. Its objective is to promote research and collaboration in mathematics and computer-based modelling.

With the increasing sophistication and power of desktop computers and the ability to run large mesh generating packages on them there has been an upsurge in interest in Applied Surface Modelling. The need for a detailed description of surfaces is becoming increasingly important, be it a fluid flow through a complicated geometry or fluid flow past an object such as a ship, plane, or car. There is also a need to consider lubrication and wear between two surfaces, and a requirement to analyze the data collected from the work. The present collection of papers attempts to address some of these problems. As the papers are presented by people from industry as well as academia the book should be of interest to a wide readership of mathematicians, computer scientists, and engineers concerned with the theories and practices relating to surface modelling.

The volume falls naturally into four sections. The first six chapters form a section on General Surface Modelling. The main theme of this section is to consider the generation of meshes to reproduce a surface. In this way a surface can be 'fed' into a computer as an algorithm. Clearly, with more and more powerful computers becoming available much more sophisticated meshes can be generated.

The next two chapters deal with the Lubrication and Wear of Surfaces. All surfaces are liable to erosion of some form. Further, if two surfaces are designed to move over one another the erosion process must be minimized, so lubrication becomes important.

The following five chapters consider the Geometric Modelling of Surfaces. In this section the main emphasis is on the reproduction of surfaces rather than the generation of meshes.

The final section on Statistical Modelling of Surfaces considers the problems associated with the accuracy and reliability of data.

The editors would like to make this opportunity to thank all the participants of POLYMODEL XII for making the conference a success. The editors must also thank in particular Gordon Moir, Terry Wilkinson, and Don Catley for their help in organizing the conference and Kathryn Oliver for her help in the administration tasks.

Newcastle upon Tyne Polytechnic — C. F. M. Creasy
July 1989 — C. Craggs

1

Commutative surprises in curve and surface theory

Malcolm Sabin
Fegs Ltd

We are used to familiar operators like sum and product being commutative, giving the same answer in whichever sequence the operands are taken. The matrix product is typically presented, when it is first encountered in a course, as something rather odd because the result of multiplication of two matrices, A and B say, depends on which comes first.

AB does not equal BA.

However, combination of operations is usually non-commutative, and any case where two operators do commute should be regarded as a special case, a reason should be sought, and the circumstances in which it occurs tightly defined.

This chapter addresses some situations in curve and surface theory where operators do in fact commute, but where the reason is not obvious. The examples are not new: they were all known by the mid-seventies, but I have not seen them looked at in this light before.

THE INTERPOLATING SPLINE

In the late 1960s the bicubic was a new discovery, and papers were published describing how an array of such patches could be fitted together into a surface, characterized by the corner points, the two tangent vectors, and the twist vectors common to all the patches meeting at a common vertex. I had the problem of building a sculptured surface system, and I required a way of calculating the various vectors from the points alone. Also relatively new was the concept of a parametric spline curve, then viewed as a way of passing a piecewise cubic curve through an ordered set of points. The theory was presented to me in structural terms.

Suppose that you have a set of y-values known at (for the moment) equal intervals of x, and all lying close to the x-axis.

A good approximation to the strain energy of a piece of springy wood constrained to pass through the known points is the integral of the square of the second derivative. This can be minimized by the calculus of variations, which says that the minimum energy curve is formed from cubic pieces, meeting at the given points with continuity of first and second derivative.

Once we know that, we can build on Ferguson's method, in which each piece of cubic is characterized by its initial and final ordinates ($y(0)$ and $y(1)$) and its initial and final first derivatives (slopes) ($y'(0)$ and $y'(1)$) (Ferguson 1964). The problem becomes how to determine the slopes.

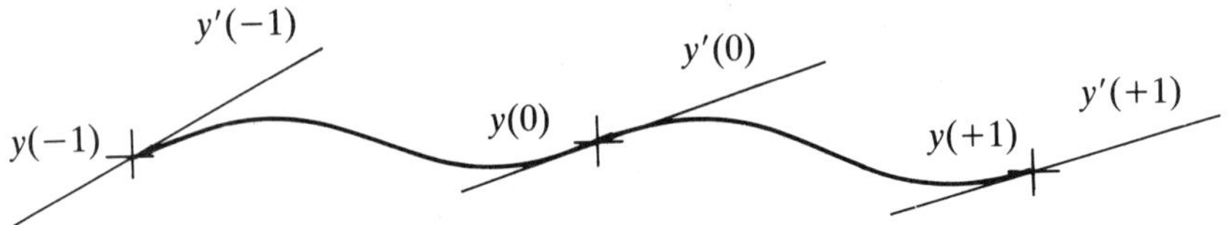

Straightforward algebra determines the actual second derivative at the left-hand end of a cubic span to be a specific linear combination of the four variables

$$y''(0+) = 6y(1) - 6y(0) - 4y'(0) - 2y'(1)$$

Similarly, the second derivative at the right-hand end of the adjacent span is

$$y''(0-) = 6y(-1) - 6y(0) + 4y'(0) + 2y'(-1)$$

Equating these two expressions gives a linear equation relating the unknown slopes to the known ordinates.

$$6y(1) - 6y(-1) = 2y'(1) + 8y'(0) + 2y'(-1)$$

This equation actually holds at all the internal data points, and so a system of simultaneous equations can be set up

$$\begin{bmatrix} \ddots & & \\ -6 & 0 & 6 \\ & & \ddots \end{bmatrix} \begin{bmatrix} \\ \mathrm{y} \\ \\ \end{bmatrix} = \begin{bmatrix} \ddots & & \\ 2 & 8 & 2 \\ & & \ddots \end{bmatrix} \begin{bmatrix} \\ y' \\ \\ \end{bmatrix}$$

or $Sy = By'$ for short

In practical terms, this is best solved by forming the left-hand side and then forwards eliminating and back substituting. However, in formal terms we may express the solution as

$$y' = B^{-1}Sy$$

Today, we would approach the whole problem quite differently. The curve would be viewed as a cubic B-spline curve whose control points were to be chosen such that the entire curve would interpolate the given data points.

Let the control ordinates by $b(i)$. From the values of the cubic B-spline basis functions at the knots we can write the interpolation equation as

$$6y(0) = b(-1) + 4b(0) + b(1)$$

$$\begin{bmatrix} \ddots & & \\ & 6 & \\ & & \ddots \end{bmatrix} \begin{bmatrix} \\ y \\ \\ \end{bmatrix} = \begin{bmatrix} \ddots & \ddots & \\ 1 & 4 & 1 \\ & \ddots & \ddots \end{bmatrix} \begin{bmatrix} \\ b \\ \\ \end{bmatrix}$$

Similarly, we can express the slopes at the data points in terms of the control ordinates

$$6y'(0) = 3b(1) - 3b(-1)$$

$$\begin{bmatrix} \ddots & & \\ & 6 & \\ & & \ddots \end{bmatrix} \begin{bmatrix} \\ y' \\ \\ \end{bmatrix} = \begin{bmatrix} \ddots & \ddots & \\ -3 & 0 & 3 \\ & \ddots & \ddots \end{bmatrix} \begin{bmatrix} \\ b \\ \\ \end{bmatrix}$$

Some trivial manipulation eliminating b now gives the form

$$y' = SB^{-1}y$$

Since this gives the same curve as the previous derivation whatever the data set the two methods are applied to, we must have the result

$$SB^{-1} = B^{-1}S$$

so that B and S are commutative matrices.

The question 'why ?' is posed, at this point, as an exercise to the reader, but properties of S and B which are easily determined are:

(i) S is antisymmetric,
(ii) the rows of S all sum to zero,
(iii) B is symmetric,
(iv) with sensible scaling (divide everything by 12), the rows of B all sum to unity. This scaling does not, of course, upset property (ii).

It may be helpful, in the search for the reason, to sharpen up on the generality of the result.

Although the outline above used the equal interval case for clarity of exposition, the same result holds for the more general case of unequal intervals of abscissa. I believe it also still holds in the case where tension parameters are included, which use the looser condition

$$y''(0-) = y''(0+) + t^* y'(0)$$

to define the slopes. This retains continuity of second derivative in the vector valued case. See Manning (1973) for the Hermite approach to this and Boehm (1987) for the B-spline approach.

The matrices S and B then become tridiagonal matrices in which each row depends on the lengths of the two intervals abutting a data point, and on the value of t at that point. B remains symmetric, and S has antisymmetry apart from the diagonal, which is not zero.

The rows of S always add to zero. The rows of B add to unity (with appropriate scaling throughout) in the equal interval case, but in the unequal interval and tension cases one can keep either the symmetry of B or the row sum property. Not both.

The real reason

In fact, the property required for commutativity is that S and B share a set of eigenvectors, and differ only in their eigenvalues. Some understanding of this comes from considering an infinitely long equal interval spline.

In this case, giving the points to be interpolated data values varying sinusoidally with distance along the spline results in both the B-spline coefficients and the first derivatives varying with the same spatial frequency. The B-spline coefficients vary in phase, and the slopes in quadrature, so B has real eigenvalues and S imaginary.

Express B and S in terms of left and right eigenmatrices L and R, and diagonal eigenvalue matrices b and s.

$$B = LbR \quad S = LsR$$

$$B^{-1} = Lb^{-1}R$$

$$B^{-1}S = Lb^{-1}RLsR = Lb^{-1}sR$$

$$SB^{-1} = LsRLb^{-1}R = Lsb^{-1}R$$

However, diagonal matrices always commute, because their product is just a term by term product of the scalars on the diagonals, and so the commutation of S and B is traceable back to commutation of scalar multiplication.

TWO WAY SPLINES

The first process described above for computing slopes at data points led me to view the spline as a differentiation operator, which was applied to a vector of points to get a corresponding vector of first derivatives. When we were given an array of points this operator could first be applied to all the rows in one direction to get the derivatives in that direction, then to all the rows in the other to get the derivatives that way.

Call the process of splining in the u-direction SU, and that in the v-direction SV.

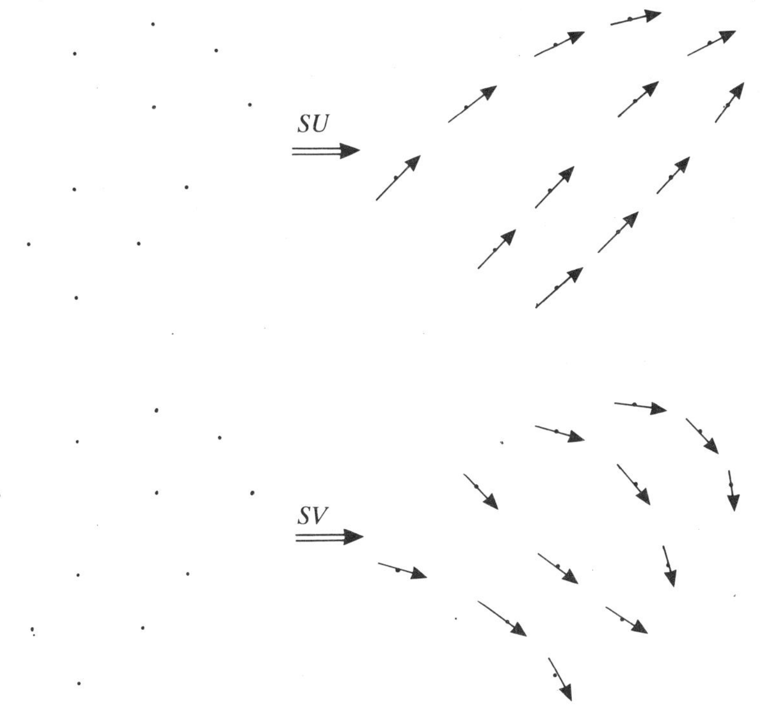

To define a bicubic patch fully, however, we needed the twist vector too, and nobody really understood its geometry then. We needed the Bezier view of the world to make twist vectors really intuitive, and so it was anybody's guess then how they should be computed.

Interpreting the spline as a differentiation operator suggested that the values of dP/dv could be splined in the u direction to give $d2P/dudv$. Or alternatively, dP/du could be splined in the v-direction to give $d2P/dvdu$.

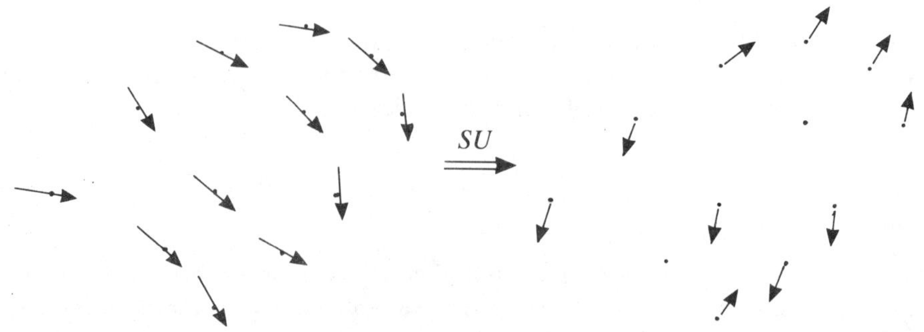

Note that I don't make the commutativity assumption yet that

$$d2P/dudv = d2P/dvdu$$

as in fact that assumption does not always hold when we deal with piecewise surfaces.

Consider the surface

$z := \text{if } x >= y$
$\quad \text{then } 0$
$\quad \text{else } (x - y)^2$
$\quad \textit{fi}$

Using the limit definition of a derivative, we can evaluate expressions for the derivatives

$$dz/dx(0,0) = 0 \qquad d2z/dydx(0,0) = 0$$

$$dz/dy(0,0) = -2(x - y) \qquad d2z/dxdy(0,0) = -2$$

This behaviour is directly related to the presence of a discontinuity of second derivative at the point where it occurs, the bicubic patches sewn together Furguson-wise can have discontinuities of second derivative.

However, each piece of surface does have a unique second cross derivative at each vertex, which is used as a handle for the characterization of the whole surface, and the question is what the value should be set to.

Our expectation was that two different processes for estimating that value would give different answers, and indeed our first implementation computed both and took the mean, not having any good argument for taking one or the other.

In fact we found that, in our circumstances, the two results were identical within computational rounding errors, and our second implementation saved itself some CPU by using only one route.

So, again we found commutativity. Representing splining with respect to u by the operator $SU(P)$, and splining with respect to v by $SV(P)$, we had found that

$$SU(SV(P)) = SV(SU(P))$$

Our codes at the time used regular grids, but the same property still holds when unequal interval grids are used, provided that the intervals are the same for all rows in the same direction. This was a limitation we were happy to accept since this is a sufficient condition for continuity of first derivative between bicubic patches, which we certainly wanted to maintain.

In fact this commutation is not limited to the spline differentiator which leads to continuity of second derivative everywhere. The property is shared by all linear differentiation operators which use the same coefficients for all those in the same direction. In particular there was some argument at the time about the merit of using the Overhauser slopes, which made each patch purely locally defined. Overhauser slope patches have first derivative continuity but not second, but the two twist estimates are still the same.

This time the explanation is straightforward, and can be outlined quite briefly.

In the modern view what we did would now be called the fitting of a tensor product *B*-spline surface. If we had used Overhauser it would still have been a tensor product surface. Now a tensor product surface has a basis function associated with each vertex. The vertices are arranged in a regular rectangular array in parameter space, and the function associated with the vertex at (i,j) is just the product of two univariate basis functions

$$f_{ij}(u, v) = g_i(u)^* h_j(v)$$

Thus the variability with respect to u and that with respect to v are separated out into the g and h functions. d/du is applied to the g part and d/dv to the h part of each term, in the overall sum. This independence means that the linear combinations used in the u-spline operator are independent of the number of times the v-spline has been applied, so giving commutativity.

COONS AND BÉZIER

Strictly speaking, my final major example is not of commutativity, but of a morphism. None the less, it is fun, because it relates theory contributed by two of the major figures in our subject, and so I include it.

The concept of a Bézier curve is a familiar one. An ordered set of points, the control points, often joined by a polygon to render the ordering visible, are the coefficients of a curve whose shape is related to the positions of the points in some very elegant ways.

For the moment forget the detail, and just regard the Bézier mathematics as a way of getting from an ordered set of points to a continuous curve. Call this mapping *Bc* (for Bézier curve).

Bc: polygon $\rightarrow$ curve

Similarly there is a bit of mathematics which gets from a regular network of points to a Bézier surface. Call this *Bs* (for Bézier surface).

Bs: network $\rightarrow$ surface

Bs is just the tensor product of two *Bc*'s, and so the two maps are closely related as well as sharing the same name.

Somewhat earlier, Steve Coons had described his blending patch, which takes a set of four curves, regarded as the four sides of a patch, and fills in the interior with a surface meeting all four boundaries however different their definitions. His first order equation is a transfinite interpolant, which is the solution of a hyperbolic differential equation.

The ability to interpolate between boundaries of completely different forms is truly marvellous, and I have always felt that calling bicubics 'Coons patches', as they have been, does neither Steve Coons nor Jim Ferguson justice. However, we can still apply Coons algorithm to curves which happen all to be generated from Bézier polygons.

Call the map from four curves to a surface *Cc* (for Coons continuous).

Cc: curve$^4 \rightarrow$ surface

In fact, the computation of the position of any specific point in the interior of the patch depends only on the corner positions and on one point on each of the four sides. This means that we can apply it equally to discrete curves, defined only for specific values of abscissae.

Call this discrete version, which maps from four polygons to a network *Cd* (for Coons discrete).

Cd: polygon$^4 \rightarrow$ network

The surprise is that for any set of four compatible polygons, X,

$$Cc(Bc(X)) = Bs(Cd(X))$$

Starting from a set of four compatible polygons, we can either generate their Bézier curves and then use Coons to put a surface in between, or else use Coons to generate a network and then apply Bézier to get the same surface.

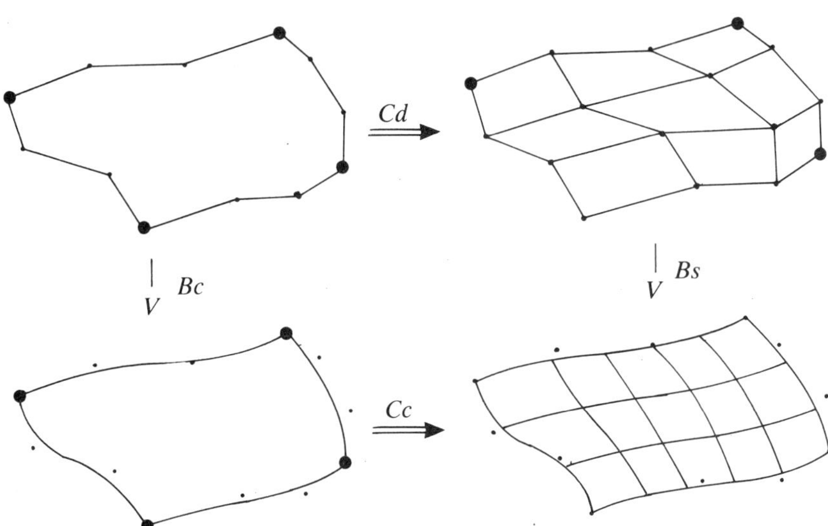

The generality of which this is a special case has been explored by Farin (see Farin 1988).

OTHER EXAMPLES

Long, long ago, the standard methods for surface representation were much more *ad hoc* than is now the case. We had methods which were tightly tied to the coordinate system, so that rotating the basic data would actually alter the shape of the surface defined. The property of coordinate system independence, which was claimed, rightly, as a major advantage of the 'new' methods, is in fact a commutativity property, that applying a rotation to the original data and then interrogating it gives the same result as interrogating first and then rotating the result. (OK, again it's strictly a morphism).

Another package, widely used, had the noncommutative property that the shape only became explicit as sections and intersections were calculated, and the final shape could depend on the sequence in which those sections and intersections were requested.

Even better known is an example from graphics. Before the meeting held by IFIP WG 5.2 at Seillac, most graphics packages were plotter oriented, in the sense that they were commanded at the move and draw level. At the conference it was pointed out how badly defined this was, in that nobody was precise about what happened if you moved the origin while the pen was down. In fact this was only one example of poor definition of the graphics pipeline, which most display manufacturers got wrong, but the conference latched on to it, and invented the GKS polyline, a concept which is now more or less universal in graphics packages.

Morphisms are inescapable in modern graphics. Clipping purports to remove those points of a line lying outside some clipping box, and this is most naturally specified in terms of operations on all the individual pixels. However, actually implementing it this way, post-interpolation, would be unacceptably slow for a software implementation. Indeed, it may not be possible if the display has hardware interpolation, and so it becomes necessary to implement the clipping by a software process operating on the pre-interpolation data, endpoints or interesting problems with thick lines and sub-pixel precision of the endpoints of the clipped line (Bresenham 1990).

The subject of finite element mesh generation is also just now in a very *ad hoc* phase, and there are a number of morphisms which can be used to challenge any scheme, and thereby possibly improve it.

Does mesh generation commute with solid body transformation of the body to be meshed?

Does mesh generation of a slightly deformed body always give a mesh only slightly deformed?

Does mesh generation of a body deformed only locally always give a mesh deformed only locally?

DOES COMMUTATIVITY MATTER

System building is becoming more and more mathematical as computers grow, together with our ambitions to fill them. More and more precise specification is becoming necessary in order for the bits of a system to fit together. The commutativity question, applied to every pair of functions in a product, is one which can be used to probe a specification, to find out if it really has defined what the software has to do.

REFERENCES

Boehm, W. (1987) Rational geometric splines *Computer Aided Geometric Design* **4** (1–2) 67–78 July 1987.

Bresenham, J. (1990) Attribute considerations in raster graphics. In: *Computer graphics theory and practice*. ed. D. F. Rogers & R. Earnshaw, Springer-Verlag (in press).

Coons, S. A. (1967) *Surfaces for computer aided design of space forms*. MIT project MAC, MAC-TR-41 (June 1967).

Farin, G. (1988) Section 20.2–20.5, pp. 268–273 in *Curves and Surfaces for Computer Aided Geometric Design*. Academic press.

Ferguson, J. (1964) Multivariable curve interpolation *J,ACM* **11** (2) 221–228.

Manning, J. (1973) Continuity conditions for spline curves *Computer Journal* **17** (2) (Feb. 1973).

2

A 3-D finite element mesh generator

D. G. Shipley
ICI Chemicals & Polymers Limited, Catalysis Research Centre, Billingham

INTRODUCTION

The Finite Element Method (FEM) is firmly established as a technique for solving problems of stress, heat flow, etc. The analysis can be split up into three distinct parts.

(1) Dissecting the structure into simple elements with the corners, and often the mid-side nodes, numbered. For some shapes this is easy but time consuming, while other shapes, the intersection of a cylinder and a torus, for example, it can entail a lot of calculation.
(2) Solving the equations.
(3) Displaying the results of the calculations.

The purveyors of software to perform (2) do not always do (1) and (3) very effectively. The program which is described below was written after the purchase by ICI of a finite element package which came with a user-hostile mesh generator. The package was to be used to analyze stresses in pressure vessels and similar structures, so a mesh generator was required for surfaces and not for solid objects. The input format described below could also be used for solid objects by defining all the surfaces: the dissection into tetrahedra would not be a very difficult programming exercise.

The author had previously written a 2-D mesh generator which required the user to specify the minimum number of nodes required to define the shape. It soon became obvious that this type of approach would not work in three dimensions because a surface cannot be defined merely by the location of its perimeter. It was therefore necessary to start from scratch and to devise a 'user-friendly' method of transferring a picture from the user's mind to the tables of elements in the computer's store. The principles adopted were:

(1) Ease of use was of paramount importance. The stress analysis section has a large workload and cannot afford to spend a lot of time generating meshes.
(2) The amount of data to be input by the user should be the minimum required to define the structure.
(3) The user should not need to know anything about node numbers or think where elements should go.
(4) The mesh generated should satisfy the demands of the stress analysis program in terms of legality of element shape, but 'beauty' was not important. People often demand aesthetically pleasing meshes, but, unfortunately, it is difficult to teach aesthetics to a computer.

THE PROGRAM AS SEEN BY THE USER

The significant features are:

(1) The user communicates with the program, using simple mnemonic commands of which the first four letters are significant. These follow the → prompt in the example.
(2) Structures are described by names of up to five characters chosen by the user.

In the current version there are 57 commands, some of which are trivial. The significant commands will be illustrated with reference to a specific example shown in Fig. 1. It consists of a circular to square transition piece with a circular pipe

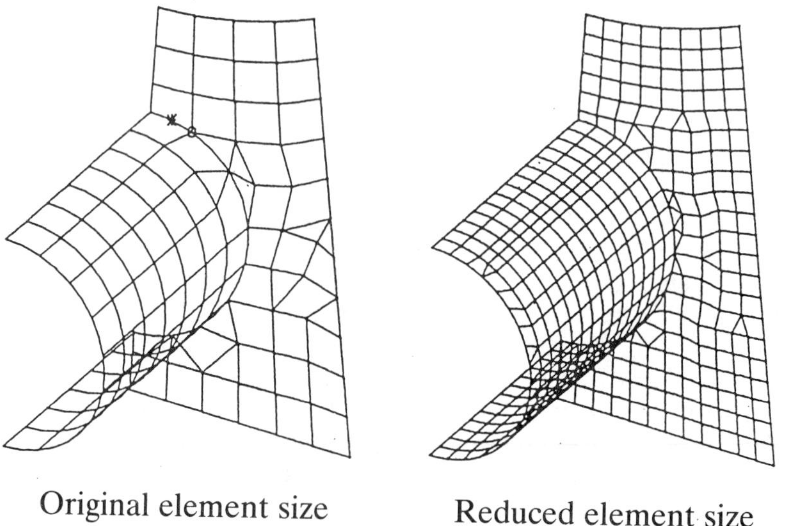

Fig. 1 — The mesh used in the example.

intersecting the transition. All commands entered, apart from display commands, are saved in a file, so that the user can restart after an error without having to re-enter the whole structure. To show how the program appears to the user, selected parts of the conversation between the user and the program are reproduced below in full. This is followed by the relevant part of the 'Save File', which contains the rest of the data entered by the user. The items in parentheses are defaults which have been added by the program to make it easier to follow. The Y's and N's are answers to questions about constraints or the retention of edge markers. When running the program the previous stucture is assumed when no name is given.

The first operation is to MAKE a simple structure. In this case it is a line, which is always made on the *X*-axis. Other options are a rectangle, a cylinder, a sphere, a torus, or a curved line passing through specified points.

This line and another are then moved to their final locations, and a flange is made by FILLing between the two lines.

An arc is then made in the *X–Y* plane, as are all 2-D structures, and moved to form the top of the cone, which is made by FILLing between the two lines.

The vertical pipe is made by GROWing from the arc, and the pieces are JOINed together to form half of one side of the duct. This is then REFLected in the *X–Z* plane and ROTated about the *Z*-axis to form the other half of the quadrant being generated.

The hole for the branch pipe is CHOPped out by a circular drill which comes down the *Z*-axis so the structure has to be rotated and moved to the required position. CHOP can also cut along any of the principal planes. The edge of the hole is saved as a structure called EDGE, and the branch pipe is made by GROWing from the edge.

The pieces are then JOINed together and returned to the vertical position to give the mesh shown on the left in Fig. 1. To produce the finer mesh shown on the right it was only necessary to change the mesh size in the Save File and re-run the program.

It is possible to add constraints to structures as they are generated and to attach plate thicknesses and physical properties to structures afterwards. Only one command is necessary to write all the relevant data in the format required by the stress analysis program.

```
EXAMPLE

-> MAKE
1   DEFINED ELEMENT
2   DEFINED LINE
3   STRAIGHT LINE
4   CIRCULAR ARC
5   RECTANGLE
6   CYLINDER
7   SPHERE
8   TORUS

ENTER CODE : 3
NAME OF NEW STRUCTURE : LINE
X1 X2 : 0,5
STEP LENGTHS : 0.8
DO YOU WISH TO APPLY CONSTRAINTS ?
   6 ELEMENTS        19 NODES
CURRENT STRUCTURE : LINE
```

```
-> MOVE
ENTER SHIFTS IN X,Y,Z DIRECTIONS : 0 -5

-> FILL LINE,ARC
NAME OF NEW STRUCTURE : LINE
DO YOU WISH TO APPLY CONSTRAINTS ?
  80 ELEMENTS        266 NODES
CURRENT STRUCTURE : CONE

-> ROTX
ANGLE OF ROTATION : -90

-> CHOP
TYPES OF CUT :
  1 AT X=0
  2 AT Y=0
  3 AT Z=0
  4 BY CYLINDER ALONG Z AXIS
ENTER TYPE : 4
1  NORMAL CUT
2  REVERSE CUT
3  NOTHING REMOVED
ENTER CODE : 1
RADIUS OF CUT : 2.3
DO YOU WISH TO SAVE THE EDGE OF THE CUT (Y/N) ? Y
NAME OF NEW STRUCTURE : EDGE
DO YOU WISH TO APPLY CONSTRAINTS ?
  160 ELEMENTS        498 NODES
CURRENT STRUCTURE : EDGE
```

The remainder of the operations required are shown as they appear in the Save File.

GROW (EDGE)
PIPE
10

(N)
JOIN PIPE CONE
(Y)
MOVE (PIPE)
0 3.5
ROTX (PIPE)
90
STOP

INSIDE THE PROGRAM

Each structure is stored as a collection of elements. The geometrical data are held on four levels of array, each one containing pointers to the level below.

(1) A table of names of structures and pointers to the beginning and end of the structure in:

(2) A list of all the elements in current use (LISTE).
(3) Tables of the numbers of nodes and node numbers in each element.
(4) The table of node coordinates.

The order in which nodes and elements appear in the lists is unimportant except that the structure currently being worked on must be at the end of LISTE so that new elements may be added. Elements are deleted merely by replacing the element number in LISTE by a zero. There is a garbage collector which is activated after any operation which could have deleted elements. This removes redundant elements and nodes from the lists and renumbers those still in use., This operation is not seen by the user, who knows nothing about node numbers.

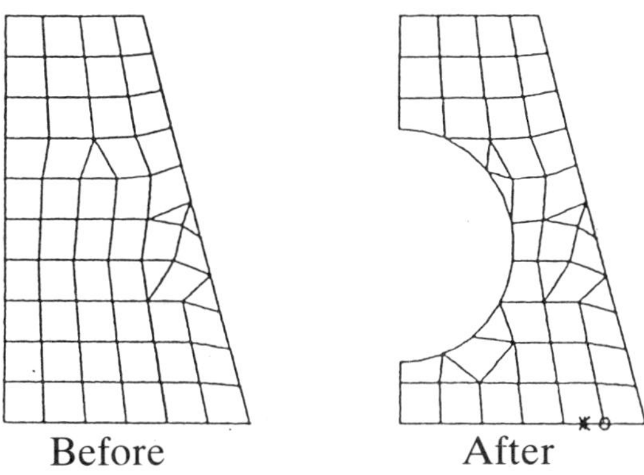

Fig. 2 — Example of a 'CHOP'.

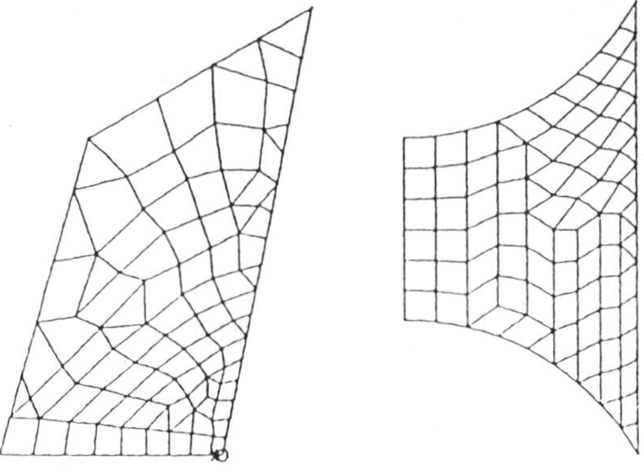

Fig. 3 — Examples of 'FILL'.

The main program reads a command and calls one of the high level routines which in turn make use of lower level routines. The routines which were the most difficult to develop were CHOP and FILL because these position nodes and aesthetic considerations caused a lot of trouble. CHOP cuts the structure and thus requires nodes to be moved to the edge of the cut or new nodes generated. Fig. 2 shows a before and after example. FILL generates elements between two lines ands can handle graded elements. Fig. 3 shows two examples of the product of this routine.

3

The generation of triangular meshes on surfaces

J. Peiro, J. Peraire and **K. Morgan**
Department of Civil Engineering, University College, Swansea SA2 8PP

We present a method for the generation of unstructured triangular meshes on surfaces. The approach affords the user great control over mesh size and grading. By employing standard methods of surface modelling, a region on a surface can be mathematically defined as a mapping between a domain in a two-dimensional (2-D) parameter plane and $\boldsymbol{R}^3$. The generation procedure takes advantage of this fact by generating a triangular mesh in the parameter plane in such a manner that, when transformed into the surface, the resulting mesh complies with the characteristics specified by the user.

INTRODUCTION

The surface triangulator which will be described has been developed with the primary intention of employing it as the first step of a 3D mesh generation procedure based upon a variant of the advancing front technique (Peraire *et al.* 1987, 1989). However, the approach discussed will be of more general interest, and applications to other areas such as, for instance, finite element analysis of shells or graphic display or surfaces, can also be envisaged.

The discretization of a region on a surface consists of positioning points on the surface and defining the links to be established between a point and its neighbours. Therefore, any surface generation method requires: (1) an analytical definition of the surface region, and (2) a criterion to position the points on the surface and define their connectivities.

In the present method, the mathematical description of the boundary of the region to be meshed is accomplished by means of parametric cubic spline curves, and the surface where the region is defined is represented by tensor-product surfaces (Faux & Pratt 1979). The meshes considered here are totally unstructured, i.e. the number of points connected to an interior point will vary through the domain. This contrasts with structured generators (Gordon & Hill 1973, Saltzman 1986) which are usually based on the generation of a topologically rectangular coordinate system on the surface. In our approach, the parameterization of the surface defines the region to be meshed as a mapping between a 2-D region in a parameter plane and $\boldsymbol{R}^3$. The mesh on the surface is obtained as the image of a triangulation of the region in the parameter plane. The triangular mesh is generated using the procedure described in (Peraire *et al.* 1989). The spatial distribution of mesh size and shape for this triangulation is defined in such a way that, after applying the mapping, the image mesh on the surface presents the geometrical characteristics required by the user. The ability of the proposed method to produce general surface meshes is illustrated by its application to the representation of a complete aircraft configuration.

SURFACE REPRESENTATION

Our problem is the discretization of a region R with boundary Γ defined on a surface S (Fig. 1). We will assume that the surface S is at least C^1 continuous and that the

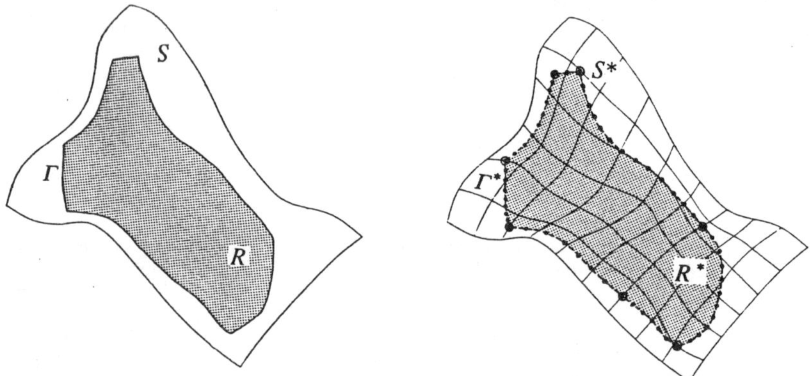

Fig. 1 — Surface representation.

boundary Γ is piecewise C^1 continuous. The boundary Γ of R will not, in general, coincide with the boundary of S.

An exact analytical definition of the surface S and the curves forming the boundary Γ is seldomly available. To represent these, we resort to the interpolation of composite surfaces and curves through a suitably defined set of points on S and Γ. These approximations will be referred to as support surfaces S^* and support curves Γ^*. In the present implementation, the support curves employed are cubic splines, and the support surfaces are represented by tensor-products of cubic splines (Faux & Pratt 1979).

SURFACE DISCRETIZATION

The above representation defines the region to be meshed R^*, with boundary Γ^*, as the image, by the mapping $x_i = x_i(u_1, u_2)$; $i = 1,2,3$ which represents the support surface S^*, of a region R, with boundary Γ, in the parameter plane, $u_1 - u_2$. If this mapping is bijective, it will transform a triangulation T in the parameter plane into a surface triangulation T* (Fig. 2). This suggests the idea of generating a mesh in the

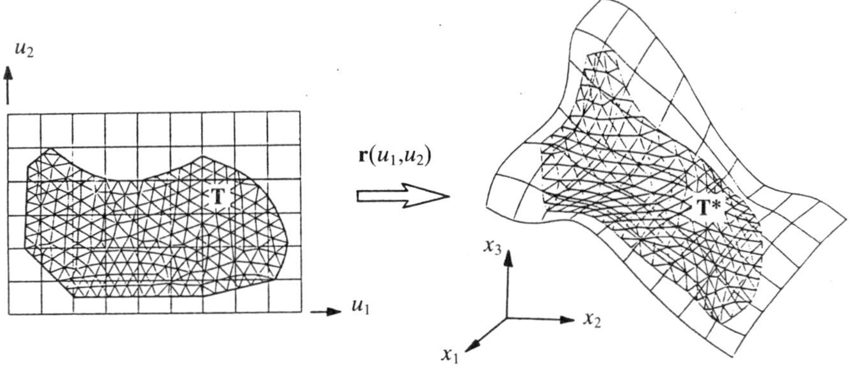

Fig. 2 — Mapping of a mesh in the parameter plane into the surface.

parameter plane which will later be mapped onto the surface to produce an appropriate surface discretization. This is accomplished by ensuring an appropriate definition of the local mesh size and shape in the parameter plane.

Mesh control: mesh parameters and background mesh

The characteristics of the triangular elements in a mesh are specified through certain mesh parameters, viz the directions of stretching $\alpha_i; i = 1, \ldots, N$ and the corresponding node spacing in these directions δ_i; $i = 1, \ldots, N$ (Fig. 3). Here N denotes the

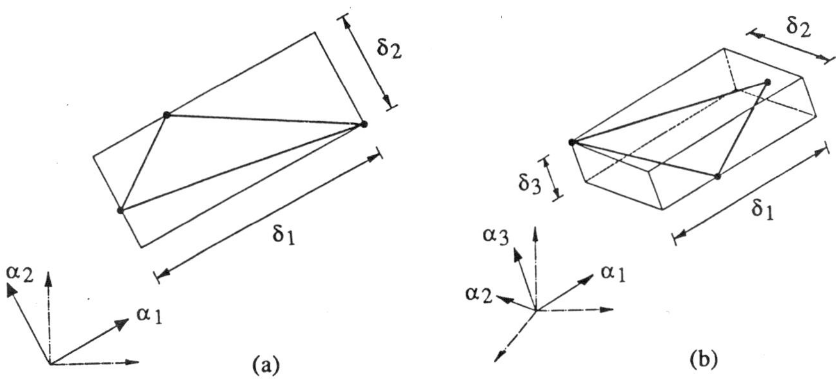

Fig. 3 — Mesh parameters: (a) parameter plane and (b) surface.

number of dimensions. The node spacing δ_i is defined as the maximum distance between the projections onto the direction α_i of the nodes of the triangle. The control over the mesh characteristics of the surface triangulation is obtained by defining a spatial distribution of mesh parameters. This is accomplished by constructing a background mesh of tetrahedra which completely covers the surface and which contains mesh parameters α_i^* and δ_i^*; $i=1,2,3$ as nodal values. During the generation, at a point on the surface, local values for these quantities are obtained by linear interpolation between the nodes of the tetrahedra to which the point is interior. The values of α_i, and δ_i; $i=1,2$ at a point in the parameter plane are computed, using the metric properties of the mapping, from the values α_i^*, and δ_i^*; $i=1,2,3$ at its image point on the surface.

Mesh generation in the parameter plane

The mesh generator utilized for the triangulation of the region R in the parameter plane is based on the advancing front technique as described in (Peraire *et al.* 1987). In this procedure the generation starts with an initial front formed by straight line segments connecting consecutive boundary nodes. Then, one of the sides of the front is chosen as the base for a new triangle to be generated. The size and shape of the triangle are taken to be the values of the mesh parameters at the middle point of the side. Depending on the geometrical configuration at hand, the new element is generated by either creating a new point or using an existing point in the front. The front is subsequently updated by deleting the sides of the triangle that were originally in the front and including those sides that were not. The process of creation of an element and updating of the front is repeated until the number of sides in front is reduced to zero. This marks the end of the triangulation.

General strategy for surface discretization

The overall surface generation strategy is conditioned by the use of the above generator. The procedural steps followed here are:

(1) definition of a suitable distribution of mesh parameters by means of a background mesh of tetrahedra,
(2) discretization of the support curves defining the boundary Γ^*,
(3) computation of the coordinates (u_1, u_2) in the parameter plane of the points generated in step (2),
(4) formation of the initial front in the parameter plane,
(5) generation of an appropriate triangulation in the parameter plane, using the 2-D mesh generation procedure,
(6) mapping of the generated mesh onto the surface.

Curve discretization

The generation process starts with the discretization of the boundary Γ^*. Points are generated on each support curve according to the specified spatial distribution of mesh parameters. Firstly, we compute the length L of the support curve. The mesh parameters are then interpolated at sampling points on the curve, and the associated spacings δ_c are computed according to

$$\delta_c = \|\mathbf{T}\, \boldsymbol{t}\|^{-1} \tag{1}$$

where $\boldsymbol{t}$ denotes the tangent to the curve at the sampling point, $\|.\|$ denotes the Euclidean norm, and **T** is an auxiliary transformation (Faux & Pratt 1979) that locally maps the 3-D space into a space in which the spacing is equal to unity in all directions: the unstretched space. Once this piecewise linear distribution of spacings along the curve has been calculated, the positions s_k of the nodes to be created are the solutions of the equation

$$\frac{N_c}{A_c}\int_0^{s_k} \frac{1}{\delta_c(s)}\, ds = k\ ; \qquad k = 0,\ldots,N_c \tag{2}$$

where s denotes the arc length parameter of the curve and N_c is the number of sides generated on the curve and which is chosen to be the nearest integer value to

$$A_c = \int_0^{L} \frac{1}{\delta_c(s)}\, ds \tag{3}$$

Computation of the coordinates (u_1, u_2) of a point

Owing to the manner in which the support curves are defined, the points generated on Γ^* are not exactly on the surface S^*, although they are very close. In this formulation, the coordinates u_1, u_2 of such a point, denoted by $\boldsymbol{r}^*$, are taken to be those of the point $\boldsymbol{r} = (x_1, x_2, x_3)$ in the surface which is the closest to it. This can be formulated as a minimization problem, i.e. find u_1, u_2 such that

$$\|\boldsymbol{r}^* - [\boldsymbol{r}(u_1, u_2)\| = \text{minimum} \tag{4}$$

The solution of (4) is accomplished by means of a simple steepest-descent-like method (Acton 1970).

It should be pointed out that the discretization of the boundary Γ^* is performed directly in the 3D space and not in the parameter plane in order to ensure compatibility between the discretized edges of surface regions when multicomponent surfaces are generated.

Mesh characteristics in the parameter plane

The spatial distribution of mesh parameters which define the geometrical characteristics of the surface triangulation T* is known in $\boldsymbol{R}^3$ through the background mesh. To construct the triangulation T, the 3D mesh parameters δ_i, α_i; $i = 1,\ldots,3$ need to be transformed to the parameter plane. By employing a modified mapping

$$\boldsymbol{R}(u_1, u_2) = \mathbf{T}(\delta_i, \alpha_i)\ \mathbf{r}(u_1, u_2) \tag{5}$$

which defines a bijective transformation between the parameter plane and the unstretched space, one can calculate the spacing δ along a certain direction $\beta = \beta_1, \beta_2)$ as

$$\delta = \left\{ \sum_{i,j=1}^{2} \frac{\partial \boldsymbol{R}}{\partial u_i} \frac{\partial \mathbf{R}}{\partial u_j} \beta_i\ \beta_j \right\}^{-\frac{1}{2}} \tag{6}$$

The 2-D mesh parameters are the directions in which, according to (6), the spacing δ is an extremum and the spacings associated to those directions. It can be easily verified that this reduces to an eigenvalue problem involving the first fundamental form of the mapping $\boldsymbol{R}$.

It is interesting to remark that, although the representation adopted for the support curves and surface provides global second order continuity, this generation procedure requires them to have only C^1 continuity.

EXAMPLES

Two examples, and representative generation timings, are presented which show the ability of the method to produce surface triangulations of variable size and shape for general geometries.

The first example is geometrically simple and is included to fully demonstrate the concepts of mesh spacing and stretching which have been introduced above. The surface used is taken in the form of half a cylinder, for which the support surface is defined by an array of 8*7 points, as shown in Fig. 4(a). The background mesh of tetrahedra is depicted in Fig. 4(b) and consists of 96 elements and 50 points. The surface meshes shown in Fig. 4(c)–4(f) were all generated by this background mesh, but employing different nodal values for the mesh parameters in each case: Fig. 4(c) shows a mesh with constant spacing and no stretching; in Fig. 4.(d) the mesh has constant stretching and a uniform spacing; a mesh with variable spacing and constant stretching has been generated in Fig. 4(e); the mesh of Fig. 4(f) has constant spacing and variable stretching.

The second example is in the form of a multicomponent surface which exhibits severe geometrical complexity and represents the geometry of an F-18 fighter configuration. From symmetry considerations, only one half of the surface need be meshed, and the surface definition is then given in terms of the 37 components and 87 intersection lines shown in Fig. 5(a). Two different views of a resulting full surface triangulation are shown in Fig. 5(b), in which the engine ducts are separately displayed, and Fig. 5(c). The mesh shown consists of 61 468 elements and 30 732 points.

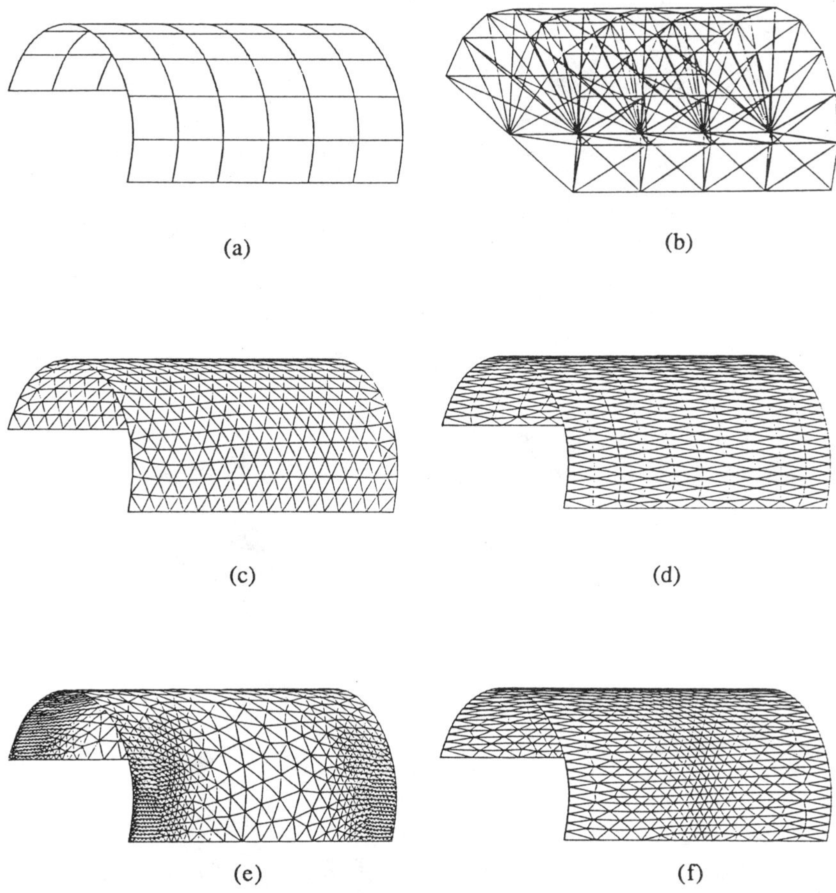

(a) (b)

(c) (d)

(e) (f)

Fig. 4 — Cylindrical surface.

The numerical performance of the proposed algorithm is illustrated in Fig. 6, which shows the variation of the cpu time, on a VAX 8700, with the number of generated surface points, for a fixed background grid. From these timings it can be seen that, for example, the generation of a mesh of the type shown in Fig. 5(b) can be accomplished in less than 1 hour.

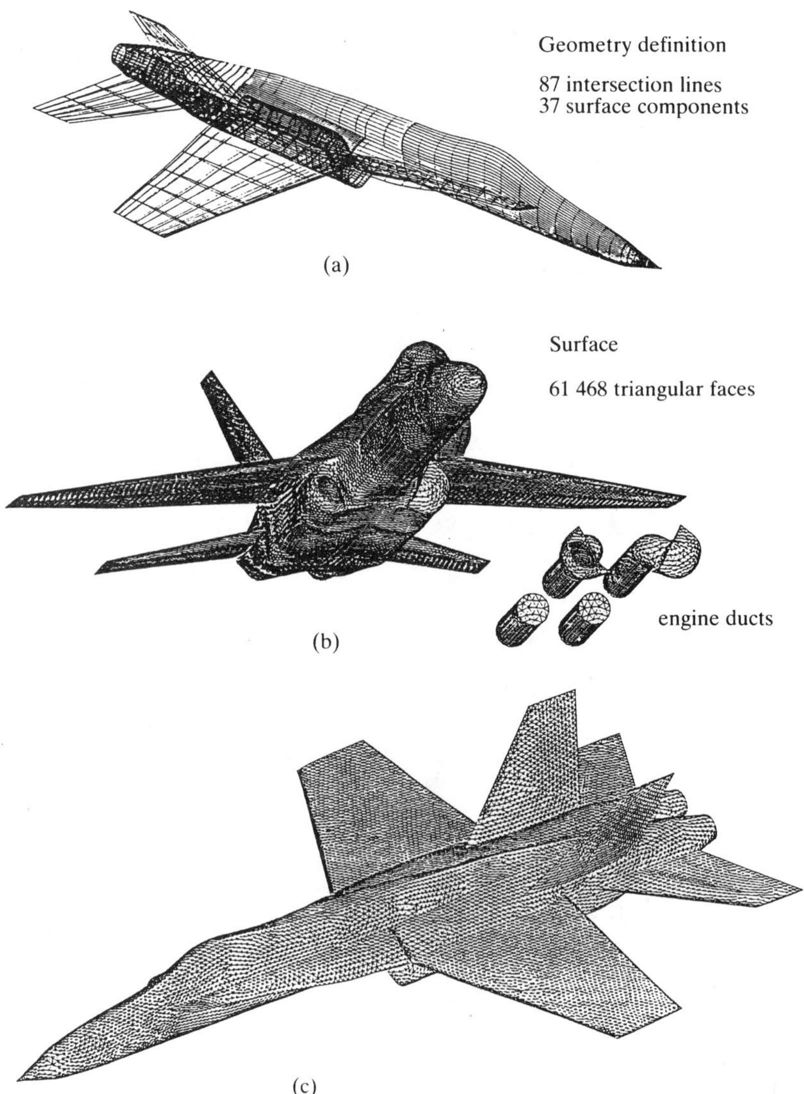

Fig. 5 — F-18 fighter configuration.

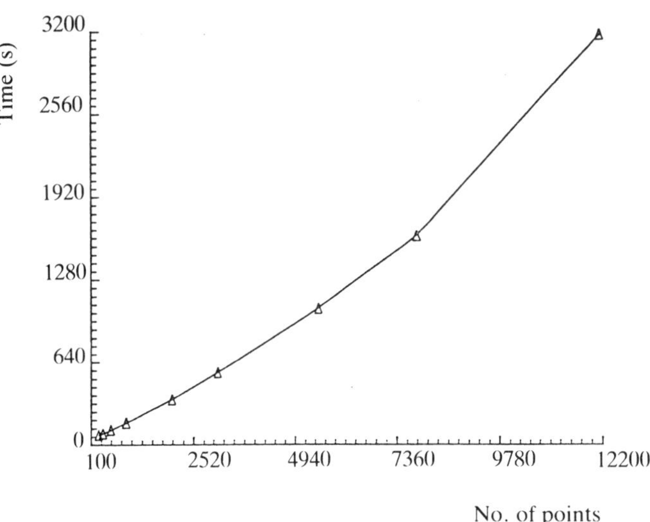

Fig. 6 — Generation timings.

REFERENCES

Acton, F. S. (1970) *Numerical methods that work*, Harper & Row, New York.

Faux, I. D. & Pratt, M. J. (1979) *Computational geometry for design and manufacture*, Ellis Horwood.

Gordon, W. J. & Hall, C. A. (1973) Construction of curvilinear coordinates systems and applications to mesh generation, *Int. J. Num. Meth. Engng.* **7** 461–477.

Peraire, J., Peiro, J., Formaggia, L., Morgan, K. & Zienkiewicz, O. C. (1989) Finite element Euler computations in three dimensions, *Int. J. Num. Meth. Engng.* **26** 2135–2159.

Peraire, J., Vahdati, M., Morgan, K. & Zienkiewicz, O. C. (1987) Adaptive remeshing for compressible flow computations, *J. Comp. Phys.* **72** 449–466.

Saltzman, J. (1986) Variational methods for generating meshes on surfaces in three dimensions, *J. Comp. Phys.* **63** 1–19.

4

Design of *n*-spline patches using recursive subdivision

D. J. T. Storry
Department of Mathematical Sciences, Leicester Polytechnic

R. S. Davies
CADCentre Ltd, Cambridge

Traditionally, the design of surface patches of other than four sides has been a difficult and time-consuming problem. In this chapter the authors describe a technique for designing such patches based on recursive subdivision, and illustrate its implementation within the GNC surface modeller.

Surface data are supplied by the user in the form of point, tangent, and 'twist' vectors at each patch corner on the surface. Rectangular patches have the usual biparametric Hermite definition. Data at the corners of a non-rectangular area is transformed into a *B*-spline net to which an optimized subdivision algorithm is applied, producing an *n*-sided *B*-spline, or *n*-spline, surface patch. The subdivision process is invisible to the user.

The resulting *n*-spline patch is itself C^2 continuous everywhere except at the so-called extraordinary point where its continuity properties are optimal. There is tangent plane continuity with neighbouring patches, as on the remainder of the surface, and one degree of freedom remains for the designer to control patch fullness.

The technique is both fast and flexible; interrogation and navigation procedures are consistent with those used over the rest of the surface. The application of the method to some common blend surfaces in engineering is illustrated.

INTRODUCTION

It is often necessary in the modelling of surfaces in engineering to incorporate patches of other than four sides. The time and effort spent in the design of such patches, however, is regularly disproportionate to the area they occupy or their importance as a feature. Moreover, it may be impossible to carry out the design within the database of the surface modeller employed

Although a number of techniques exist for the design of *n*-sided patches (Boehm *et al.* 1984), those most commonly used, such as degeneracy of patch representation (Bézier 1974) or kinked parametrization of patch assembly (Bézier 1974, Storry & Ball, submitted) are unnatural in their mathematical form and may place unwelcome restrictions on the design of the surrounding rectangular areas. The approach described here, based on recursive subdivision, overcomes both these problems.

RECURSIVE SUBDIVISION

The use of recursive subdivision for the design of surfaces of arbitrary topology was first suggested by Catmull & Clark (1978). They demonstrate that a uniform parametric bicubic B-spline surface over a rectangular patch carpet can be regarded as the limit of an infinite process of subdivision applied to the corresponding rectangular mesh of control points. They describe a generalization of this algorithm to a polyhedral mesh of arbitrary topology, producing in the limit a continuous surface. After the first iteration the mesh is entirely rectangular, except at a fixed number of so-called extraordinary points, corresponding to vertices where other than four edges meet or to nonrectangular faces in the original configuration. Fig. 1

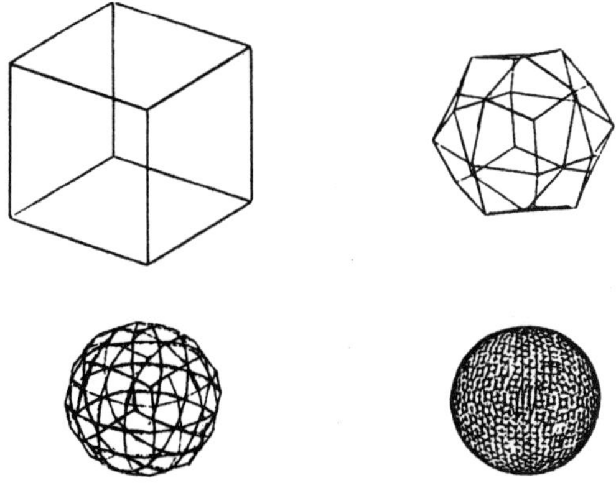

Fig. 1 — Recursive subdivision applied to a cube.

shows an example of such a subdivision procedure applied to a cube.

We consider the algorithm in its most general form. Fig. 2 shows the effect of one

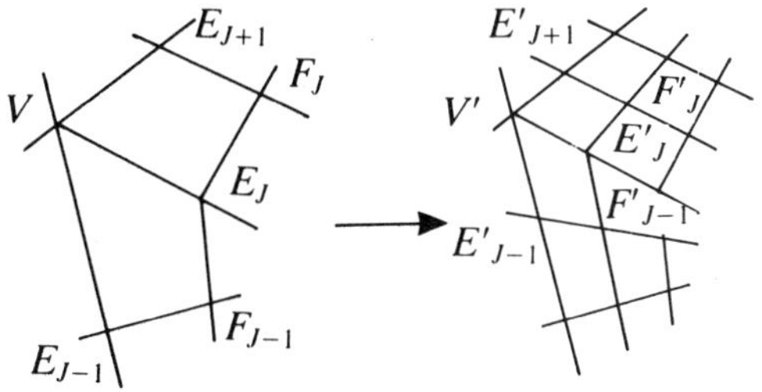

Fig. 2 — The effect of one subdivision.

subdivision at a point V in the mesh where n rectangular faces meet. For $j = 1, 2, \ldots, n$, V is connected by an edge to the point E_j and shares a face with the point F_j. The subdivision algorithm is given as follows:

$$\left.\begin{aligned} &V' = \alpha_n V + \beta_n\left(\frac{1}{n}\sum_{i=1}^{n} E_i\right) + \gamma_n\left(\frac{1}{n}\sum_{i=1}^{n} F_i\right) \\ &\text{where } \alpha_n, \beta_n, \gamma_n \geqslant 0 \quad \text{and} \quad \alpha_n + \beta_n + \gamma_n = 1. \\ &E_j' = \tfrac{3}{8}(V + E_j) + \tfrac{1}{16}(E_{j-1} + F_{j-1} + F_j + E_{j+1}) \\ &F_j' = \tfrac{1}{4}(V + E_j + F_j + E_{j+1}) \end{aligned}\right\} \quad (1)$$

In order that a rectangular mesh yields the corresponding B-spline surface, then necessarily

$$\alpha_4 = \tfrac{9}{16} \qquad \beta_4 = \tfrac{3}{8} \qquad \gamma_4 = \frac{1}{16}$$

Consequently, for an arbitrary mesh, the limit surface of the rectangular area is simply the corresponding B-spline patch carpet and is thus C^2 continuous. It is only at the limiting positions of extraordinary points where problems in smoothness may arise.

CONTINUITY PROPERTIES AT EXTRAORDINARY POINTS

While Catmull & Clark judged the quality of subdivision surfaces in empirical terms, Doo & Sabin (1978) related their continuity properties to the eigenproperties of some specific variants of the subdivision transformation. Ball & Storry (1986) adopted a totally general approach and derived the spectrum of the transformation given in (1). The results may be summarized as follows:

(a) 1 is an eigenvalue

(b) Corresponding to the two degrees of freedom in the subdivision weightings α_n, β_n, γ_n, there are two eigenvalues λ_a, λ_b, given by the roots of the quadratic

$$q_0(\lambda) = \lambda^2 - (\alpha_n - \tfrac{1}{4})\lambda + (\tfrac{1}{16} - \tfrac{1}{8}\beta_n)$$

By convention $|\lambda_a| \geqslant |\lambda_b|$

(c) The remainder of the eigenvalues, the so-called natural spectrum, depend only on n and occur in pairs λ_w, λ_w' ($w = 1, 2, \ldots, n$) given by the roots of the quadratic

$$q_w(\lambda) = \lambda^2 - (\tfrac{5}{8} + \tfrac{1}{8}c_w)\lambda + \tfrac{1}{16}$$

where $c_w = \cos(2w\pi/n)$

In a subsequent paper Ball & Storry (1988) established a condition on these eigenvalues so that the surface at the extraordinary point is C^1 continuous. Tangent plane continuity is attained if and only if $|\lambda_a| < \lambda_1$, that is, the larger eigenvalue dependent on the subdivision weightings has modulus strictly less than the largest eigenvalue in the natural spectrum.

It can be shown that unless $n = 4$ it is in general not possible to achieve C^2 continuity at the extraordinary point (Ball & Storry, submitted). However, by choosing $\lambda_a = \lambda_1^2$, disparities in curvature on curves through the extraordinary point can be minimized. For $n = 3$, the most common example in practice, this disparity is less than 10%. For $n > 4$, the phenomenon described by Doo & Sabin as 'discontinuity of fractional derivative' is observed, and so ratios of curvature must be treated circumspectly. For $n = 5$, another common case, the discrepancy is likely to be less than 5%, and in any event at least as good as for $n = 3$. For n even, non-trivial C^2 continuity may be achieved in certain circumstances.

DESIGNING THE *n*-SPLINE FROM HERMITE DATA

Consideration may now be given to the problem of filling in an n-sided hole R in an otherwise locally rectangular surface as shown in Fig. 3. For $j = 1, 2, \ldots, n$, R adjoins

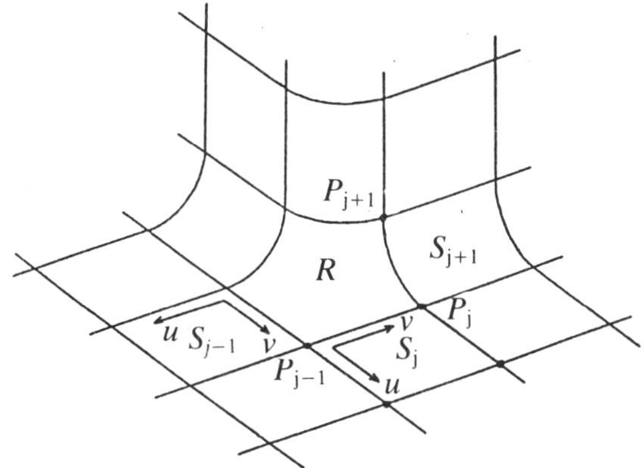

Fig. 3 — Configuration of patches S_j around an isolated 5-sided area R showing the orientation of parametrization.

a parametric bicubic patch $\{S_j(u, v); 0<u, v<1\}$ the orientation of whose parametrization is shown. It is assumed that there is parametric C^1 continuity between the rectangular patches and that they can be represented in Hermite form, consistent with the definition of such patches in the GNC surface modeller. The patch corners adjacent to R are labelled P_j $(j = 1, 2, \ldots, n)$.

Then

$$
\begin{aligned}
P_{j-1} &= S_j(0,0) = S_{j-1}(0,1) \\
P^u_{j-1} &= S^u_j(0,0) = S^v_{j-1}(0,1) \\
P^v_{j-1} &= S^v_j(0,0) = -S^u_{j-1}(0,1) \\
P^{uv}_{j-1} &= S^{uv}_j(0,0) = -S^{uv}_{j-1}(0,1)
\end{aligned}
$$

where the superscript denotes partial differentiation.

Now a *B*-spline surface arising from the subdivision process can be regarded as an assembly of rectangular patches containing *n*-sided regions which correspond to the area around extraordinary points in the original mesh. Storry & Ball (submitted) have shown that given the *n*-sided hole R and its corner points, tangents and 'twists', $P_j, P^u_j, P^v_j, P^{uv}_j$ respectively, an *n*-sided *B*-spline patch or *n*-spline can be fitted into R with continuity of position and tangent plane over the boundary.

The technique entails identifying the following points around an extraordinary point V for $j = 1, 2, \ldots, n$, as shown in Fig. 4.

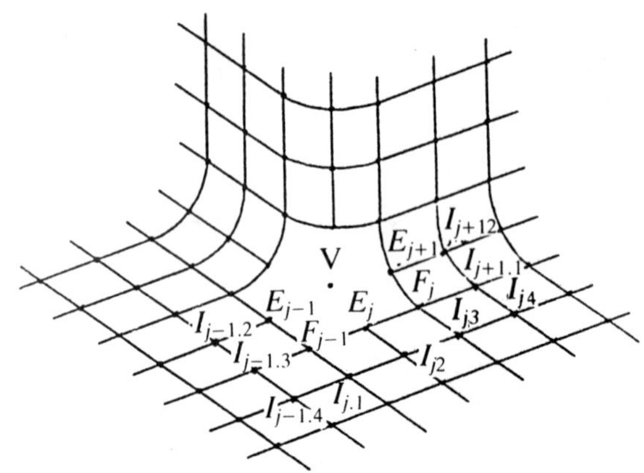

Fig. 4 — Configuration of *B*-spline points.

E_j connected by an edge to V
F_j sharing a face with V
I_{jk} $(k = 1, 2, 3, 4)$ a further ring of points.

Necessary and sufficient conditions can then be established so that the *n*-spline surface resulting from the application of the subdivision procedure to the mesh formed by these points interpolates the boundary of R with the required continuity properties. In this way E_j, F_j and I_{jk} may be expressed in terms of the P_i, P_i^u, P_i^v, P_i^{uv}, the surface descriptors required by the modeller. One degree of freedon remains in the choice of these points, and considerations of symmetry suggest this is taken up by the extraordinary point V. This corresponds to giving the designer the facility to specify the degree of fullness of the patch by his choice of patch centre.

The resulting surface now has C^1 continuity over all boundaries, and, within the *n*-spline itself, C^2 continuity everywhere except at the extraordinary point. Assuming the subdivision algorithms is optimized as described in the preceding section, there is at least tangent plane continuity here.

The subdivision process itself is invisible to the user. Point, tangent, and 'twist' vectors for the corners of R are inherited from the definition of the surrounding patches, and the desired patch centre either supplied by the user or given a default value. Routines in the modeller now compute the set of points

$$\{V\} \; u \; \{E_j, F_j, I_{j1}, I_{j2}, I_{j3}, I_{j4}\}_{j=1,2,\ldots,n}$$

and perform the optimized subdivision algorithm. The resulting *n*-spline may now be interrogated in a fashion similar to the remainder of the surface.

PRACTICAL CONSIDERATIONS

When a designer comes to design a surface, the first task is to arrange a network of control curves on the surface. With most current surface modellers there is considerable flexibility in designing the shape of each curve, but there are some constraints on their assembly. In particular, the patches enclosed by the curves of the network must be 4-sided as shown in Fig. 5.

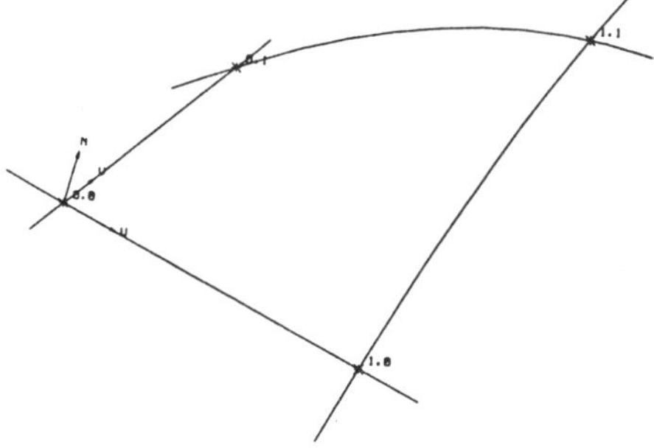

Fig. 5 — Rectangular patch formed by curve network.

There are many examples of surfaces where it is simply not possible to arrange this kind of rectangular network, or it can be achieved only with considerable distortion of the net and an equal amount of ingenuity on the part of the designer.

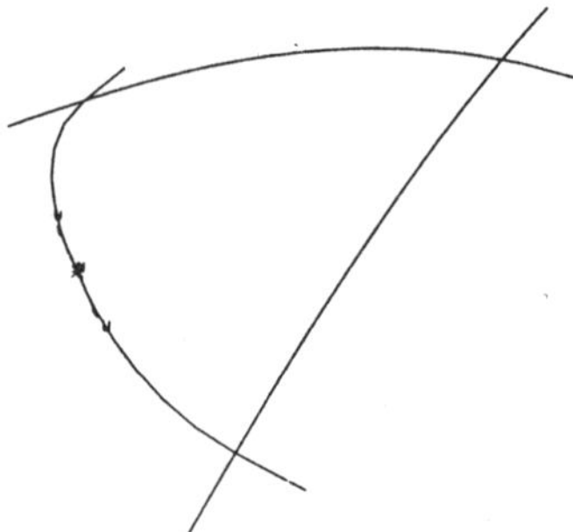

Fig. 6 — Degenerate 4-sided patch.

Fig. 6 shows one method of designing a naturally triangular region with a 4-sided patch. Two of the edges of the patch have been arranged so that they are tangent continuous at the ends where they meet. This method sometimes works, though a consideration of the isoparametric lines on this patch shows how highly distorted it is near the coalescing edges.

Two practical problems arise with this patch. Firstly it is impossible to continue the 4-sided network across the coalescing edges, and it is very difficult to match this boundary to a contiguous surface with tangent continuity. Secondly, the surface normal is not properly defined at the meeting of the coalescent edges, and this makes it difficult to create an offset surface for machining. The GNC surface modeller is used for both surface design and NC surface machining, and, in contrast to many surface modellers, the machining aspects are considered very important at the CADCentre.

Both of these problems are overcome with the use of the *n*-spline surface patch. A natural rectangular network can be extended from all sides of the *n*-spline patch, and the surface normal can be evaluated everywhere within it, even at the extraordinary point. Thus the offset surface can be calculated safely everywhere without distortion effects.

Figs 7 and 8 show two practical examples of surfaces which benefit from the inclusion of *n*-spline patches. Fig. 7 represents a surface which might be used in designing the wing section of a car, the shaded area showing where a 3-sided patch fits more naturally than a 4-sided one. In Fig. 8 a set of 5-sided patches have been enbedded in the bottle surface to form the blend region into the handle section. It is hard to see any other method that could be used satisfactorily to design the shape of this bottle.

Many more examples could be quoted to further illustrate the point, all of them taken from common, everyday objects.

n-SPLINE SURFACES IN GNC

In GNC the *n*-spline surface has been implemented as a single patch surface, to be used in conjuction with one or more regular surfaces. The underlying mathematics

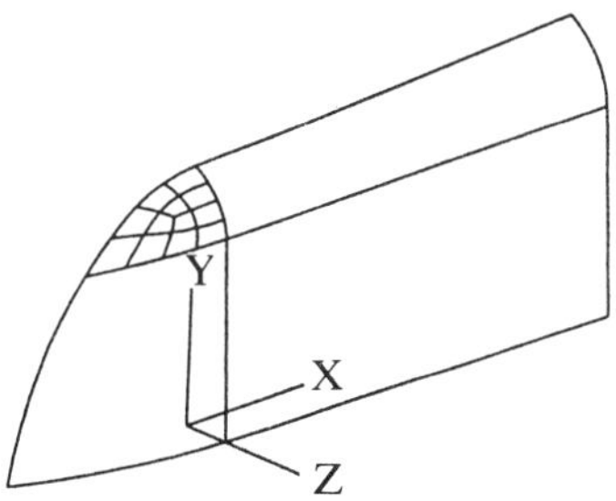

Fig. 7 — Triangular patch.

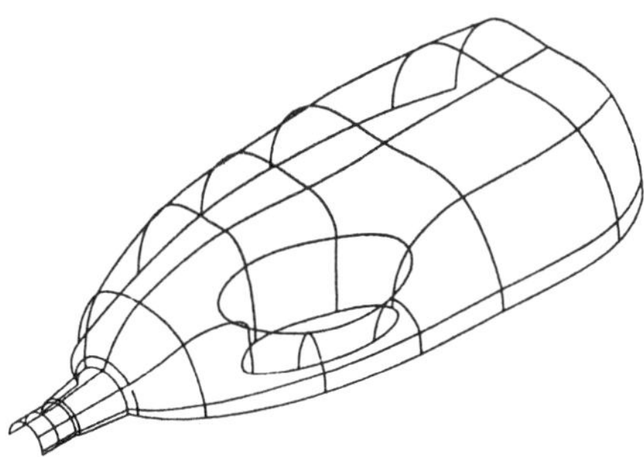

Fig. 8 — Bottle with handle region incorporating 5-sided patches.

ensures that tangent continuity between an *n*-spline and its neighbours is maintained if the *n*-spline nodes, that is the definition vectors at the patch corners, are set up correctly. The simplest way to achieve this is to set up the *n*-spline nodes directly from the nodes of the surrounding surfaces to produce a 'natural' blend surface over the irregular region. The user is still left with freedom to modify the surface locally because of the facility for independently adjusting the position of the centre point of the *n*-spline; in this way the fullness of the surface can be altered without affecting the slope continuity with adjacent surfaces.

Since there seem to be a limited number of part configurations in which irregular patches will commonly be used, the modeller has been equipped with a number of automatic generation options to cover them. These predefined cases include common situations such as rounding off corners with 3-sided patches and creating various types of pipework blends with 5- or 6-sided patches.

Fig. 9 shows an *n*-spline which has been created to fill the triangular region shown in Fig. 5. The definition vectors of the three nodes are inherited from the adjacent

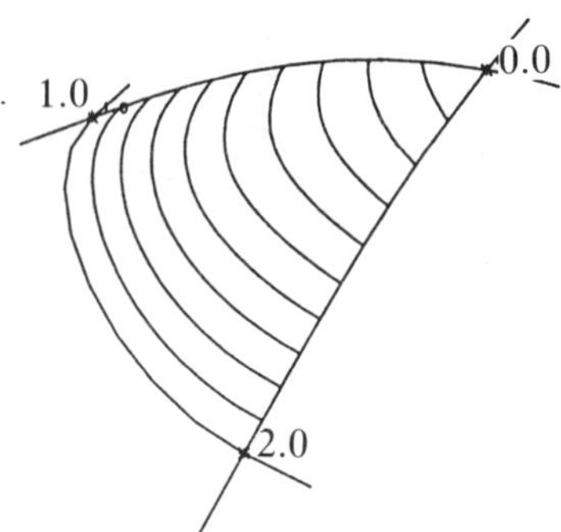

Fig. 9 — Triangular *n*-spline.

regular surfaces. Fig. 10 shows cross-sections from the offset surface; this is radically superior in quality to that obtained from the degenerate 4-sided patch.

SPECIMEN JOB

The surface in Fig. 11 was created in GNC to model an electrical space-connector, the section shown being an enlargement of one side. It can be seen that the patches in the region around the shoulder of the wide section are highly distorted. In addition, two parameter lines at the top of this section run into one. These features create severe interrogation problems in that parameter lines do not remotely correspond to cross-sectional contours.

Fig. 12 shows a preparametrization of the surface which has parameter lines lying in a much more natural fashion in relation to the part description. This new model is composed of three separate surfaces: a single regular surface which defines the overall shape of the part and two *n*-splines which fill in the shoulder region. For this situation, one 3-sided and one 5-sided patch are the most natural way of modelling the irregular region. The two *n*-splines are shown in Fig. 13 with a net of interior lines drawn to give an idea of the resulting shape.

Fig. 14 illustrates the machining paths for the composite surface. These can be broken into two separate sections. The first is an area clearance of the side wall using cutter paths generated from the parametric part of the regular surface. In this region the parameter lines correspond to the natural machining paths, and, being easy to compute, they are the simplest way of machining the side wall.

The same approach cannot be applied on the top of the surface because there is no continuous parametrization over the shoulder region where the *n*-splines have been inserted. Since it is still preferable to have continuous machining paths over the top of the part, a set of cross-section lines have been used to clear this region. Those curves form unbroken paths over the composite shape formed by the regular surface and the *n*-spline.

CONCLUSION

In general, current surface modellers are limited in the range of sculptured surfaces that they can produce by their restriction to 4-sided patches. The authors have

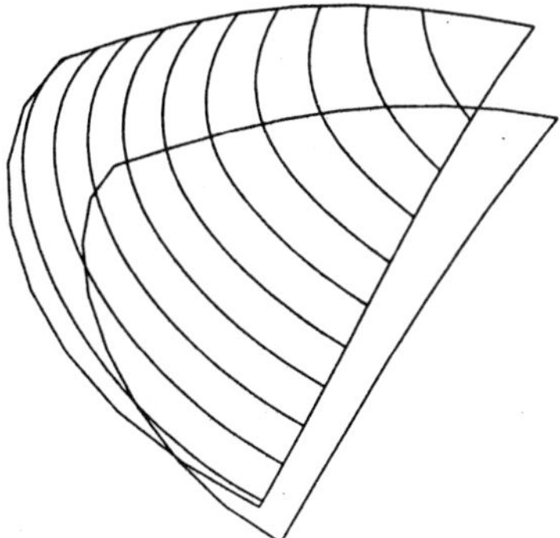

Fig. 10 — The offset surface.

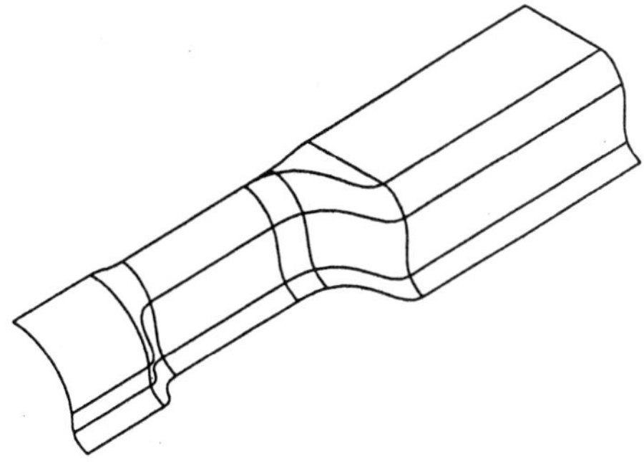

Fig. 11 — Spade-connector with distorted parametrization.

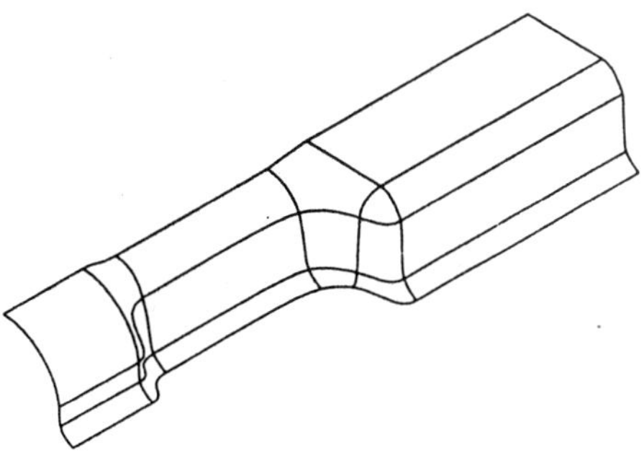

Fig. 12 — Improved patch configuration

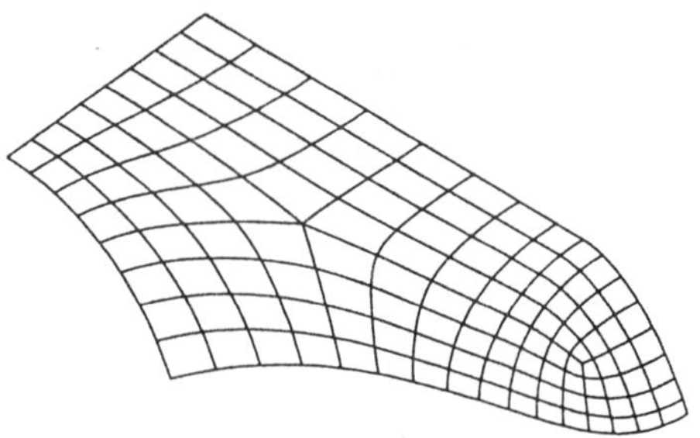

Fig. 13 — *n*-splines in the shoulder region.

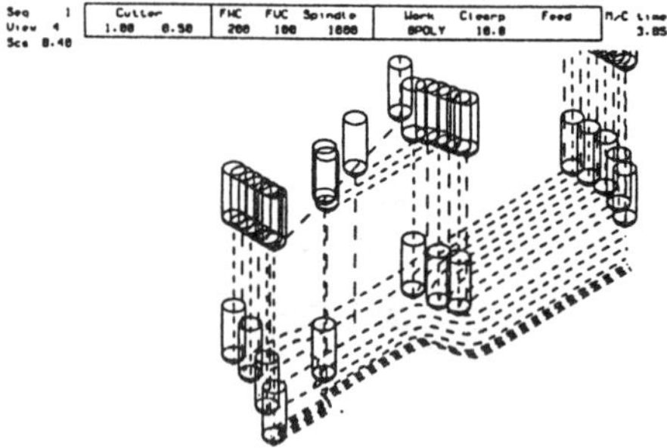

Fig. 14 — Machining paths.

attempted to describe how the technique of recursive subdivision can be used to generate patches of other than four sides, so-called *n*-splines, in a manner consistent with the design of the remaining rectangular part of the surface.

This consistency is twofold. Firstly, the surface descriptors required for *n*-spline design are just those required for the rectangular assembly and in general are inherited from neighbouring patches. Secondly, the continuity properties of the *n*-spline, both internally and across its boundaries, match those of the rest of the surface. In addition, the designer retains a degree of freedom to control the fullness of the *n*-spline shape.

This facility has been incorporated into the GNC surface modeller in such a way that, for a non-technical user, an already difficult discipline is not made more so.

Acknowledgements

The authors are pleased to acknowledge support from the Science and Engineering Research Council and the contribution of C. G. Singleton, Spaceward Microsystems, Ely.

REFERENCES

Ball, A. A. & Storry, D. J. T. (1986) A matrix approach to recursively generated *B*-spline surfaces, *Computer Aided Design* **18** 437–442.

Ball, A. A. & Storry, D. J. T. (1988) Conditions for tangent plane continuity over recursively generated *B*-spline surfaces, *ACM Transactions on Graphics* **7** 83–102.

Ball, A. A. & Storry, D. J. T. *An investigation of curvature variations over recursively generated B-spline surfaces* (submitted for publication).

Bézier, P. (1974) Mathematical and practical possibilities of UNISURF in computer-aided geometric design (R. E. Barnhill & R. F Riesenfeld, eds), Academic Press, 127–152.

Boehm, W., Farin, G. & Kahmann, J. (1984) A survey of curve and surface methods in CAGD, *Computer Aided Geometric Design* **1** 1–60.

Catmull, E. & Clark, J. (1978) Recursively generated *B*-spline surfaces on arbitrary topological meshes, *Computer Aided Design* **10** 350–355.

Doo, D. & Sabin, M. A. (1978) Behaviour of recursive division surfaces near extraordinary points, *Computer Aided Design* **10** 356–360.

Storry, D. J. T. & Ball, A. A. (1989) Design of an *n*-sided surface patch from Hermite boundary data, *Computer Aided Geometric Design* **6** 111–120.

Storry, D. J. T. & Ball, A. A., Approximation of an *n*-sided surface patch using kinked parametrization (submitted for publication).

5

Extending FAG based feature recognition

D. E. R. Clark
Department of Mathematics, Heriot-Watt University, Edinburgh
J. R. Corney
Department of Mechanical Engineering, Heriot-Watt University, Edinburgh

Expert systems can automate the evaluation of many different solutions to a problem, each of which may have its own advantages. A plan for the machining of simple $2\frac{1}{2}$-D engineering components often represents only one of many possible plans. The number of possible plans is dictated largely by the topology of an object and the capabilities of the machine tool being used to form it. The choice of a particular plan, however, is dictated largely by the geometry of the component. In this chapter we briefly review how face adjacency graphs can be used to represent the topology of an object. We then outline the key elements of a scheme for representing the capabilities of a 3-axis milling machine in terms of the changes it can make to a components topology. Finally, we demonstrate how this representation could be used to generate a graph of possible machining strategies, suitable for use by a nonlinear AI planner.

INTRODUCTION

Many complex mechanical engineering components used in the aerospace industry are formed by milling from billets (blocks) of aluminium. Because of this, although the final shape of a component may have a large number of faces it can be fully described in terms of cylindrical and planar surfaces.

Even with only a basic knowledge of machining it can be appreciated that there are many different possible ways in which final shape could be generated from the initial block. Not only are there many possible methods of manufacture, but the time taken to program the computer controlled machine tools to carry out any one particular order of cuts can take days.

The generation of these part programs would be speeded up if solid models of the component generated during its design could be automatically interpreted and passed on to an AI planning system capable of investigating the trade-offs between different strategies and determining the best plan of manufacture. Such a system is being developed under the auspices of an ACME funded project between Heriot-Watt University, Ferranti Infographics, and Ferranti Defence Ltd.

Before we can apply AI techniques to plan the manufacture of a component we must be able to represent the geometric and topological information that comprises the model in a symbolic form. In other words we must identity manufacturing features in the model. This chapter describes one approach to this problem, but *not* the method currently implemented in the project mentioned above.

BOUNDARY REPRESENTATION OF OBJECTS

Boundary representation (B-rep) describes an object by means of the surfaces which enclose it. The boundary is assumed to be partitioned into a finite number of bounded subsets called faces, where each face is, in turn, represented by its bounding edges, and each edge in its turn represented by its vertices. We define our terms as follows.

- A boundary representation of a solid object Σ is a triple $B_\Sigma = (V_\Sigma, E_\Sigma, F_\Sigma)$, in which V_Σ, E_Σ, and F_Σ denote the set of vertices, edges, and faces of Σ respectively.
  In addition to the three primary entities of face, edge, and vertex, we define secondary entities.
- A shell of an object Σ is a maximal connected set of faces of Σ. The number of shells in Σ is called the multiplicity of Σ.
- A loop on a face f of Σ is a closed chain of edges bounding f.

In a B-rep a clear separation can be made between the topological and the geometric descriptions of an object. The topological information is concerned with the connectivity (adjacency) relations between pairs of individual entities. Geometric data, on the other hand, define the shape, space location, and orientation of each primitive entity.

Most commercial B-rep modellers are based on the so-called winged-data structure first introduced by Baumgart (1975) and use of a five-level hierarchic data structure (like the one shown in Fig. 1) to represent the relationship between V_Σ, E_Σ and F_Σ.

However, although such a structure contains all the topological and geometric information relating to a body, it is designed to be robust and computationally efficient rather than to facilitate the searching and identification of patterns (which is effectively what features are) in groups of contiguous faces.

Because of this we have concentrated on a different boundary representation developed by some Italian researchers (Ansaldi *et al.* 1985, Floriani & Falicidieno 1988) in the early 80s called the Face Adjacency Graph (FAG) or Face Adjacency Hyper Graph (FAH).

FACE ADJACENCY GRAPHS

The FAG treats faces as the basic nodes of the data structure and concentrates on explicitly representing their adjacency relationships.

- The face adjacency graph (FAG) associated with an object Σ is a graph $G_\Sigma = (V, E)$ such that

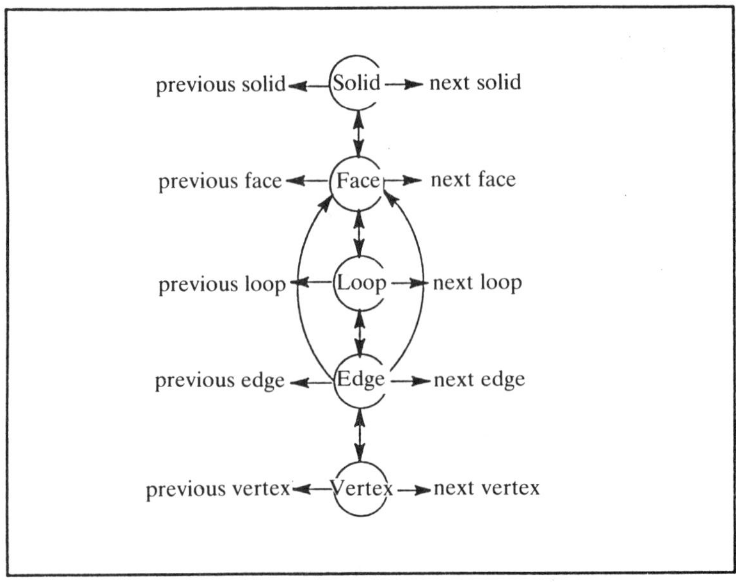

Fig. 1 — Typical winged-edge data structure.

(FAG 1) for every face f of Σ there exists a unique vertex in V corresponding to f. This vertex is labelled V_f.

(FAG 2) for every edge e, shared by two faces f_i and f_j of Σ, there exists a unique edge e_{ij} in E, connecting the two vertices of G_Σ corresponding to f_i and f_j respectively.

Since two faces may share more than one edge, G_Σ will, in general, have multiple edges. Fig. 2 shows the FAG for a cube with a pocket.

Between the vertices and edges of a B-rep we can define two relations:

- Two faces f_1 and f_2 are called edge-adjacent (resp. vertex-adjacent) if there exists an edge e in E_Σ (resp. vertex V in V_Σ) shared by f_1 and f_2.

PROPERTIES OF FAGs

It is not surprising that much of the topology of Σ may be obtained from G_Σ. We list some of these properties without proof. See (Berge 1973), Hopcroft & Tarjan (1973a,b, 1974) for details.

(1) The multiplicity of Σ is related to the components of G_Σ. The FAG of an object Σ composed of only one shell is connected, while every component of G_Σ corresponds to a shell of Σ. Hence, the multiplicity $\sigma(\Sigma)$, can be computed directly from G_Σ by invoking one of the many standard algorithms for determining the components of a graph. Fig. 2a shows a FAG of multiplicity one.

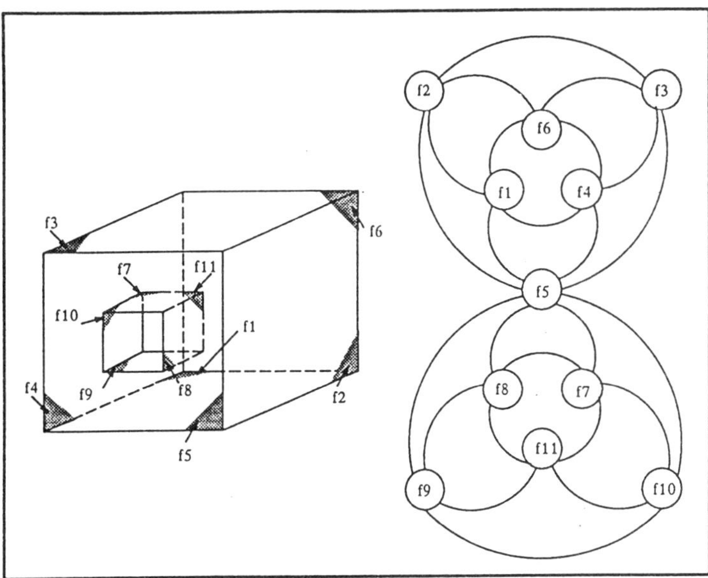

Fig. 2a — FAG for cube with pocket.

(2) The vertex-connectivity of G_Σ is related to the hole-loops and also to the genus, $g(\Sigma)$, of Σ. In particular, if $g(\Sigma) = 0$, then every component of G_Σ, is 1-connected if and only if Σ has a multiply-connected face (e.g. one containing inner loops). In this case, every multiply-connected face of Σ corresponds to a cut-vertex in G_Σ. Fig. (2a) shows a FAG with one cut-vertex labelled 'f5'.

(3) The number of biconnected components of G_Σ is equal to $h(\Sigma) + 1$ where $h(\Sigma)$ denotes the number of hole-loops of Σ. Thus, each loop on a face of Σ is defined by the vertices of G_Σ adjacent to the vertex V_f which belong to the same biconnected component. Whence the hole-loops of Σ can be computed directly by invoking one of the standard algorithms for determining the biconnected components of a graph. Fig. 2b shows a biconnected graph (i.e. removal of vertices f6 and f5 would cause the FAG to split into two components).

(4) Every handle (or hole) through Σ which defines exactly one hole-loop on two of the faces, corresponds to a separation pair in G_Σ. If on Σ any handle (or hole) through Σ defines a hole-loop on exactly two faces, and furthermore, any hole-loop on a face corresponds to a hole (or handle) on Σ, then $g(\Sigma) > 0$ if and only if Σ contains at least one separation pair (such as f6 and f5 in Fig. 2b). In this case, the number of triconnected components $\tau(\Sigma)$ of G_Σ is equal to $g(\Sigma) + 1$; and furthermore if every component is triconnected then $g(\Sigma) = 0$. Whence $g(\Sigma)$ may be computed directly by invoking one of the standard algorithms for determining the triconnected components of a graph.

(5) If we allow interior loops on the faces of Σ then every loop on a face f is defined by those faces which correspond to the vertices of G_Σ which are adjacent to vertex V_f and belong to the same triconnected component. If the objects represented by

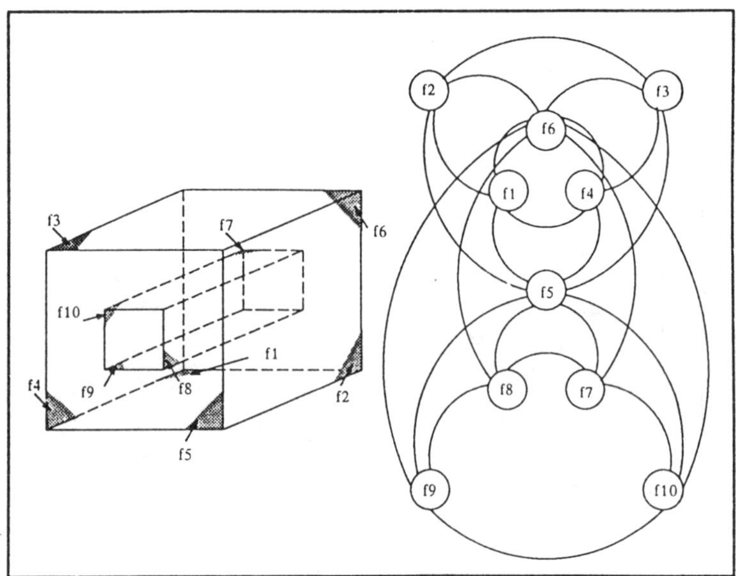

Fig. 2b — FAG for a cube with a through pocket.

the triconnected components of G_Σ have genus zero then G_Σ is embeddable on the 2-sphere. Whence the vertices of Σ may be determined from G_Σ by invoking standard algorithms for computing planar embeddings of a graph.

FACE ADJACENCY HYPERGRAPHS

We can encode more topological information directly into G_Σ to form a Face Adjacency Hypergraph (FAH), which explicitly represents both edge-adjacency and vertex adjacency relationships between the faces in Σ.

Specifically, we replace condition (FAG 2) in the definition earlier by:

(FAG 2) For every edge e in E_Σ shared by the two faces f_i and f_j, there is a unique edge in E joining the vertices V_{f_i}, and V_{f_j}. This edge is given three labels.

(i) the names of the vertices of the corresponding edge e in Σ that it represents,
(ii) the names of the two loops in Σ to which e belongs,
(iii) the angle θ between the outward normal vectors to faces f_i and f_j (positive for convex, negative for concave, see later).

We add

(FAG 3) Let $F_v = \{f_1, f_{i_2}, \ldots\ldots\ldots, f_{i_k}\} \subset F_\Sigma$ be the set of faces of Σ incident with a vertex V of V_Σ, and let $N_v = \{V_{i_1}, V_{i_2}, \ldots\ldots\ldots, V_{i_k}\} \subset V$ be the subject of vertices of G_Σ corresponding to the faces in F_v. Then, for every vertex V in V_Σ, there exists a unique hyperarc that conencts the vertices of N_v. This hyperarc is labelled $h_{i_1}, h_{i_2}, \ldots\ldots\ldots, h_{i_k}$.

Remarks

(1) Since the vertex-edge relation is encoded into the labelling of the edges of G_Σ, property 2(i) ensures that the FAH can describe objects with faces sharing two or more edges.
(2) Property 2(ii) automatically encodes the loop structure of Σ into the FAH.
(3) Property 2(iii) encodes some of the geometry of Σ into the FAH..
(4) By attaching the vertex and loop labels to its edges the FAH provides a complete topological description of any solid object bounded by a compact, oriented surface.

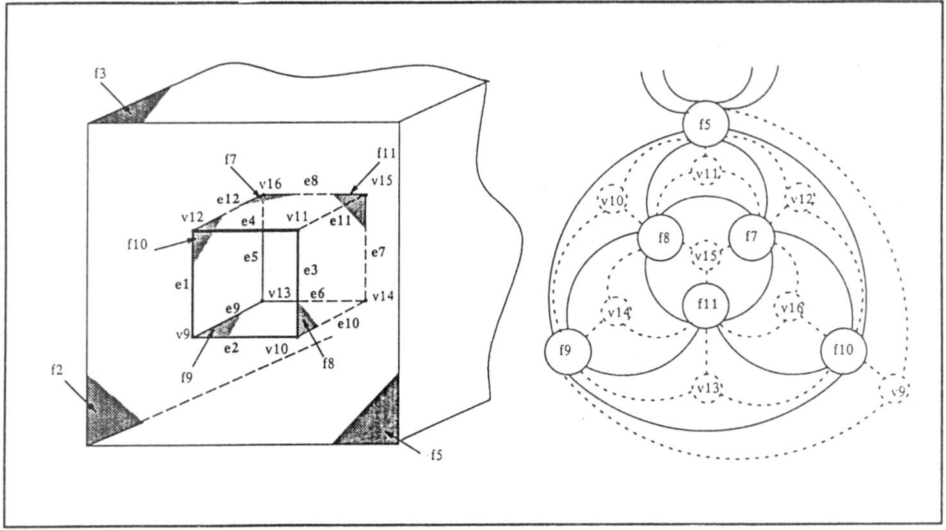

Fig. 3 — Showing part of the FAH for the object in Fig. 2a.

(5) All three primitive topological entities (vertex, edge, and face) describing the boundary of an object are represented by distinct elements of the FAH.
(6) The FAH may be disconnected. Every shell of Σ is represented by a distinct connected component of the FAH, the converse of which is also true.
(7) Since self-loops are permitted the FAH can also represent objects with curved faces (e.g. circular pockets and pillars).

Next, we impose three ordering structures on the edges, hyperarcs and hyperarc vertices:

(OI) The edges of E incident with any given vertex V_f in G_Σ are partitioned into distinct ordered sequences. Ordered according to the order of the corresponding edges on the face f of Σ represented by V_f, that is, if e_i and e_j denote two edges of E incident with vertex V_f, then e_i is an immediate predecessor of e_j⇔the corresponding edge $\tilde{e}_i$ in E_Σ is the immediate predecessor of an edge e_j in E_Σ corresponding to e_j in the loop of edges of Σ bounding face f. Each is composed of only those edges that have the same loop label — and so correspond to a distinct loop on face f.

Remark

This ensures that the edge-loop relation is encoded in the FAH. Whence all the edges bounding any given face f of Σ may be recovered from G_Σ. This is done by considering the edges of G_Σ incident with vertex V_f according to their loop labels, and then writing down the (ordered) edge sequence within each set of such edges which have the same loop label.

(OII) Organize the hyerarcs in H incident with any given vertex V_f of G_Σ into an ordered sequence. The order being that of the corresponding vertices on face f of Σ represented by V_f.

(OIII) Organize the extreme vertices of any given hyperarc $V_{i_1}, V_{i_2}, \ldots\ldots, V_{i_k}$ of G_Σ into an ordered sequence, the order being that of the corresponding faces around the vertex of Σ represented by the hyperarc $V_{i_1}, V_{i_2}, \ldots\ldots, V_{i_k}$.

INCLUSION OF GEOMETRY IN FAHs

Measures of a component's geometry can easily be incorporated into the FAH representation (Joshi & Chang 1988). Every arc can be associated with the angle formed by the two faces sharing that edge. In this chapter, edge arcs are marked with a 1 to denote an angle of $+\frac{\pi}{2}$, a convex edge or marked with a 0 to denote an angle of $-\frac{\pi}{2}$, a concave edge. Unmarked edges can be assumed to be convex. We refer to these as decorated FAH.

The present authors are experimenting with matrices that can be used to mark the nodes of FAHs.

RECOGNITION AND GENERATION OF MACHINING STRATEGIES

One of the main motivations behind the creation of FAHs is as a representation that facilitates automatic recognition of many 'features' present in a component. We can define a feature as a 'pattern of geometry and topology that is of interest'.

The FAHs give us a language in which these patterns can be defined. The FAH representation of the B-rep data structure is more amenable to feature recognition work than, say, the winged-edge, where properties such as adjacency and connectivity provide more important information than geometric data.

Different types of engineer will see different features in the same component. A stress analyst would be interested in, say, features that give rise to stress concentrations (such as sharp changes in section). A manufacturing engineer, on the other hand, would look for features that dictate or suggest certain forming processes. Because of this we feel that it is unrealistic to expect models to be labelled at creation time. The features, in other words, are not intrinsic attributes of the body.

Our particular interest is the generation of machining strategies for simple (planar and cylindrical faces only) $2\frac{1}{2}$-D bodies, so we not unnaturally wish to identify the features associated with milling operation.

We assume that the starting point of the forming process is a cuboidal billet. The problem we wish to solve can then be stated as 'how do you transform the billet into the shape of the final component using the forming operations available'?

The effects of simple milling operations on a FAH can be represented by the addition of subgraphs.

Fig. 4 illustrates how three common milling operations can be represented in this way.

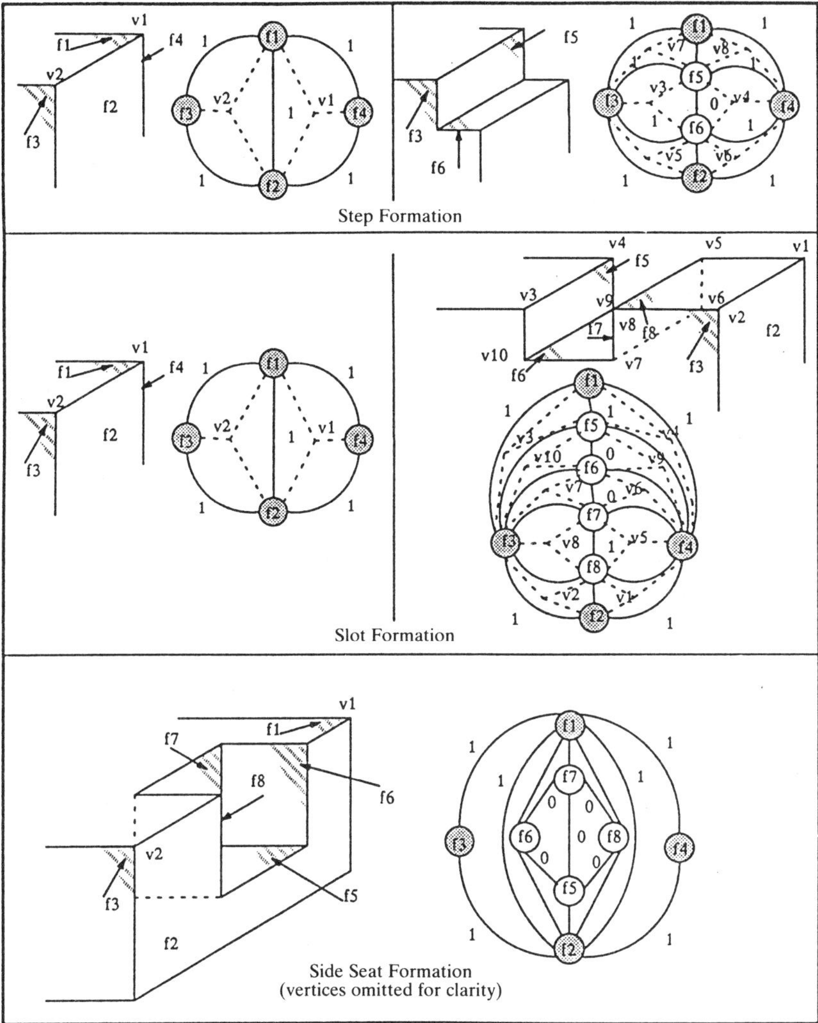

Fig. 4 — Three common milling operations represented by the addition of FAH subgraphs.

Notice that the patterns in these simple cases are easily identifiable, as are the initial conditions.

In essence we are saying that each milling operation in our idealized world can be characterized by the number and orientation of new faces, edges, and vertices it produces on a body.

Our approach is to identify and recursively remove all the machined features from a decorated FAH. Such a procedure could be used to generate a graph between the billet (Σ_i) and the final component (Σ_f) where each node represents a possible intermediate stage of the component's manufacture and each arc represents the addition or subtraction (depending on the direction the graph is traversed) of a forming operation subgraph. We call such graphs 'strategy graphs'. Fig. 5 shows one

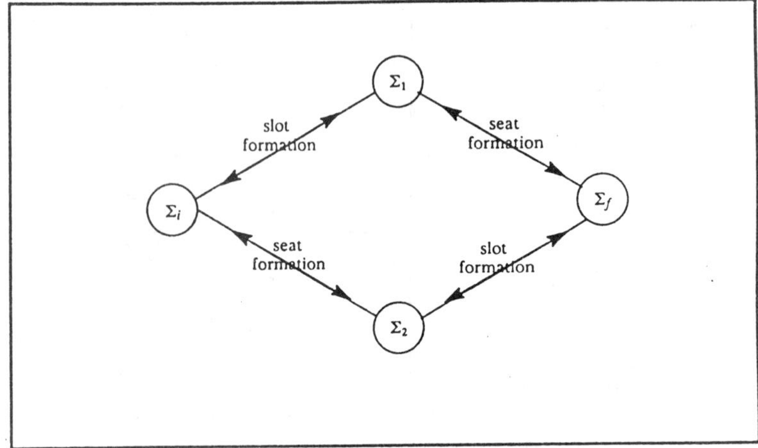

Fig. 5 — Strategy graph.

such graph, and Fig. 6 illustrates the FAGs and shapes that could be associated with one branch of it. Such a graph would represent all the possible ways a component could be manufactured, and could be used as a starting point for an AI planning system capable of making decisions based on analysis of the component's geometry.

An important point to make is that unlike other feature recognition schemes we do not attempt to produce a unique feature hierarchy or tree. This choice of graph rather than a hierarchy will allow us to represent the inherent ambiguity of many situations where features interact by sharing faces, such as the one shown in Fig. 7.

CONCLUSION

Decorated FAHs offer a succinct and flexible representation for reasoning about an object's topology. It has been shown how the effects of idealized $2\frac{1}{2}$ milling operations can be represented by the addition of subgraphs to FAHs.

Furthermore, it has been shown how the identification and removal of these feature subgraphs could yield multiple paths from billet to component.

In this chapter, however, only simple examples without feature interactions have been dealt with in a qualitative manner. It is the authors' hope that an algebra can be

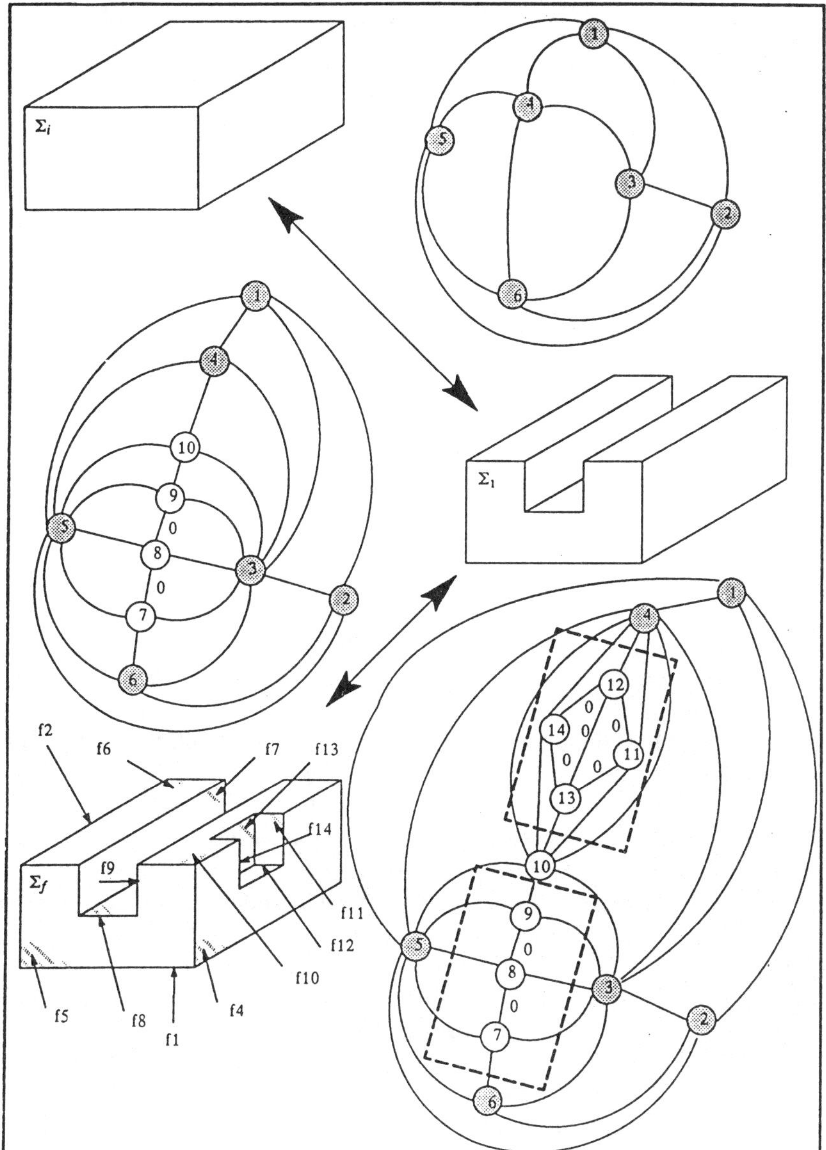

Fig. 6 — Illustration of the strategy graph shown in Fig. 5.

developed to define the effects of inserting and removing subgraphs, such as those illustrated. Such an algebra would allow the generation of strategy graphs in terms of a forming machine's capabilities.

ACKNOWLEDGEMENTS

The authors would like to thank Dr Iain Donaldson and Dave Willis for arousing our interest in face-adjacency graphs. We further wish to thank Professor Jim Murray for

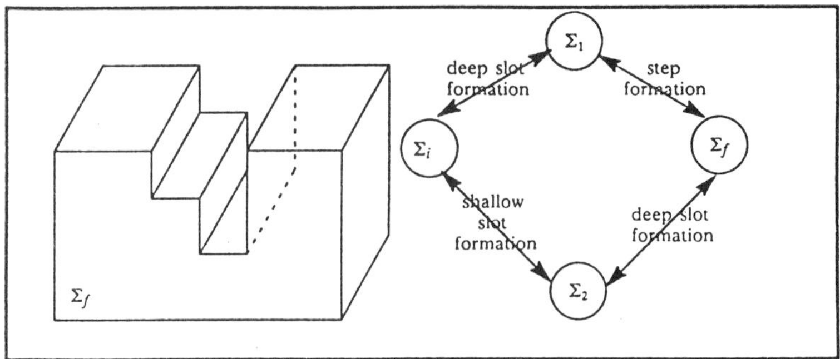

Fig. 7 — An example of interacting features.

his continual support and encouragement. We would also like to thank the ACME Directorate, Ferranti Infographics and Ferranti Defence, whose support generated our interest in FAH.

REFERENCES

Ansaldi, S., De Floriani, L. & Falcidieno, B. (1985) 'Geometric modelling of solid objects by using a face adjacency graph representation', *Comput. Graph* **19** 3 131–139, *SIGGRAPH* '85 *Conf. Proc.*, San Francisco, California, July 22–26.

Baumgart, B. (1975) 'A polyhedron representation for computer vision', *National Computer Conference*, 589–596, *AFIPS*, *Conf. Proc.*

Berge, C. (1973) *Graphs and hypergraphs*, North Holland, Amsterdam.

Deo, N. (1974) *Graph theory with applications to engineering and computer science*, Prentice-Hall.

Floriani, L. & Falicidieno, B. (1988) 'A hierarchical boundary model for solid object representation', *ACM Transactions on Graphics* **7** 1 42–60.

Harary, F. (1989) *Graph theory*, Addison-Wesley, Reading, Mass.

Hopcroft, J. E. & Tarjan, R. E. (1973a) 'Dividing a graph into triconnected components, *SIAM J. Computing* **2 3**(a) 135–158.

Hopcroft, J. E. & Tarjan, R. E. (1973b) Efficient algorithms for graph manipulation, *Comm. A.C.M.* **16** 2(b) 77–84.

Hopcroft, J. E. & Tarjan, R. E. (1974) 'Efficient planarity testing', *Journal C. M.* **31** (4) 549–568.

Joshi, S. & Chang, T. C. (1975) 'Graph-based heuristics for the recognition of machined features from a 3D solid model', *Computer-aided design* **20** (2).

Requicha, A. A. G. (1981) 'Representation of rigid solids, theory, methods and systems', *A.C.M. Computing Surveys* **12** (4) 682–690.

White, A. T. (1984) 'Graphs, groups and surfaces', North Holland.

6

Possibilities for increased availability of sculptured surface software

Tor Dokken
Center for Industrial Research, P.O. Box 124, Blindern, 0314 Oslo 3, Norway

The mathematics behind software for sculptured surface modelling is known to only a very limited number of people. For this reason, the industrial use of sculptured surfaces mathematics is less than the actual industrial needs. To simplify the introduction of sculptured surface mathematics into new applications, we have been developing subroutine libraries containing general sculptured surface functionality. These packages facilitate the building of new applications. However, the programmers still have to have detailed knowledge of the basic mathematical algorithms and formats. The introduction of standards like UNIX, X-windows, and PHIGHS facilitates the building of portable systems that can offer general sculptured surface functionality, a programming interface, MACRO possibilities, a MAC-like user interface, and access to standards for geometry transfer. Such systems will satisfy a demand for sculptured surface systems that can be tailored to solve application-oriented problems.

INTRODUCTION

Based on the resources put into research in the sculptured surface area, sculptured surface technology should be more used in the industry. In Norway the work of Even Mehlum on nonlinear splines played a central role in the Autokon systems, a CAD/CAM-system for shipbuilding, developed in the sixties and seventies. The use of this software in the automotive and aircraft industry was no great success. In the late seventies and the start of the eighties we developed at SI FORTRAN subroutine packages for sculptured surface modelling. The *B*-spline libraries supporting these packages have found a wide use. Now we are developing a spline package programmed in C. In the next section we will try to look into the problems connected to the developing and marketing of sculptured surface software. Then, in the section following, we will try to outline what the future sculptured surface tool box should contain to be a good tool for building application oriented systems. In the section following thereafter, we go into some aspects of documentation of geometric functionality and mathematical solution methods.

THE FAILURES AND SUCCESSES IN EARLIER DEVELOPMENT ATTEMPTS

To develop a good sculptured surface system requires developers familiar with software engineering, numerical mathematics, and industrial applications. To find personnel having all of these qualities, can be a great problem. During the development of the AUTOKON systems we had personnel with all of these qualifications. The AUTOKON system had a leading position in the world market for CAD/CAM systems for shipbuilding for many years. However, when the basic modelling functionality of the systems was marketed to the automotive and aircraft industry, the knowledge of these industries in the company marketing the modelling system was too limited, despite the fact that many of the basic modelling algorithms needed there are the same as in the shipbuilding industry.

In 1978 we started an Inter-Nordic CAD/CAM project called GPM (Geometric Product Models). At SI the work was concentrated around sculptured surfaces, and the software divided into two software packages: a basic *B*-spline library working only on arrays, and a boundary structure sculptured surface package utilizing the *B*-spline library. Experiences over the years show the following: in many applications the boundary structure complicated the systems being developed. In addition the number of alternative top level functions in a modelling package is enormous, and we had developed only a small selection of the alternative functions. Another important aspect is that people tend to design their own data structures; nobody likes to import data structures designed by others into their systems. The development continued in the German–Norwegian CAD/CAM- project APS (Advance Production Systems) 1981–1987. Here the activities at SI were concentrated around extending the basic functionality of the *B*-spline package. This package is currently used in a number of different CAD and FEM applications.

The distribution of the software is, however, not as wide as we originally expected. One reason is that, to develop programs based on the libraries, you need programmers with good knowledge of spline theory. We have experienced that the programmer with no background in spline theory gets an intuitive feeling about how the mathematics works. However, they will not utilize the software in an optimal way.

THE NEXT GENERATION SCULPTURED SURFACE SYSTEMS

To get a wider use of the sculptured surface software an alternative is to make a flexible mathematical application oriented toward users with a good mathematical background. Since in most cases they will not be advanced programmers, the threshold for increasing the functionality of the system should be made as low as possible.

Looking into the success of systems having a Mac-like user interface, the logical step would be to provide a user interface that is as easy to use as a system running on a Mac. Since the user dialogue in many cases will be 3-D, resources must be invested for developing new classes of interactive modelling techniques.

The basic functionality of the system should contain all necessary basic geometry operations like curve definition, surface definition, intersection algorithms, and calculation algorithms. At the same time it is vital that information can be read from and to other systems via standard interfaces like VDA-FS, IGES, and STEP.

It is obvious that such a system can never cover the wide range of functionality necessary for all users. Thus extendability of the systems is vital. Two ways should be supplied, a powerful MACRO facility and a programming interface to the basic spline algorithms. To facilitate the programming of new functions by users, the source code of the application routines must be available. Thus examples of programming are available, and functions that are variants of functions already existing, can easily be made.

Since standardization of software components is getting more important, the system should be based on, for example, C, UNIX, X-windows, and PHIGS, or alternatives to these. This will simplify the porting of the system between different types of workstation.

By introducing a system architecture as indicated, a new type of modelling tool can be supplied. The system can be tailored to the functionality needed for very special geometric operations, and information can be received from and sent to CAD/CAM-system.

DOCUMENTATION OF BASIC FUNCTIONS

In addition to the description of the functionality of functions, the mathematical description of the solution techniques should be described. An example of a function creating a number of problems is curve/curve intersection. When doing curve/curve intersection 2-D, we want to find points where the curves intersect exactly. When doing curve/curve intersection 3-D, we want to find local minima of the distance function between the two curves, where the distance is less than a specified geometric tolerance. For the mathematical oriented user the difference between these two functions is obvious. For others it is difficult to understand the difference. Thus a very detailed documentation not only of geometric operations but also of the mathematical solution strategy is necessary. It is important to introduce a mathematical way of thinking in the documentation of the system.

By putting emphasis on the documentation of the mathematical solution strategies, the user of the software will be encouraged to formulate the problems in a mathematical way. Experience shows that when developing new functionality in advanced CAD/CAM-systems the developers tend not to describe the problems in a mathematical way, thus instead of solving their original problem, they solve a related problem, with the result that the functionality of the functions developed is not according to expectations.

CONCLUSION

Hardware and software standards emerging can be used for making portable sculptured surface modelling systems. These can be extended to problem oriented modelling in a more flexible way than traditional CAD/CAM-systems. Information transfer to and from the system will be through standard geometry transfer interface formats.

REFERENCES

Bjørke, Ø. (ed.) (1987) *Advanced production systems*, Tapir Publishers, Trondheim, Norway. ISBN 82-519-0821-3.

Dokken, T. (1982) Some calculations on *B*-spline curves. *Proceedings from CAD'*82. Butterworth & Co. (Publishers) Ltd, Guildford, Surrey, England. ISBN 86-103-058-3.

Dokken, T. (1984) A design system for products with sculptured surface, *Proceedings from CAD'*84. Butterworth & Co. (Publishers) Ltd, Guildford, Surrey, England.

Dokken, T. (1986) Some applications of sculptured surface modelling, *Proceedings from ICOGRAPHICS, ETAS PERIODOCO S.p.A*, Milan, Italy 1986.

7

Surface modelling in tribology

J. A. Greenwood
Department of Engineering, University of Cambridge

If we want to describe two surfaces rubbing together, we obviously need to model the surfaces. The first model is rather boring, but a very good approximation: it looks like Fig. 1a. You may think that this might do for the surface of a brick, or a lathe

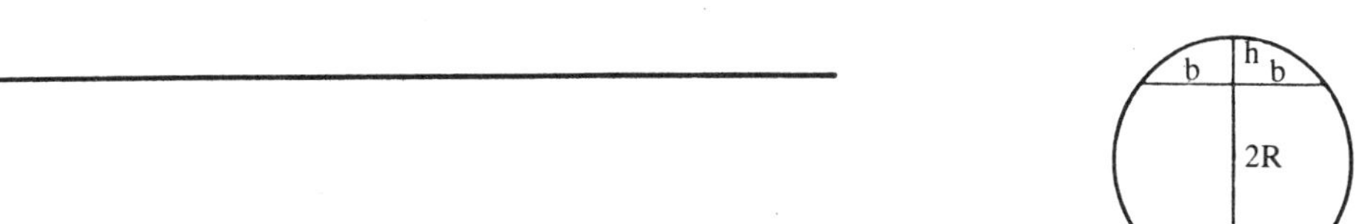

Fig. 1a Model surface

Fig. 1b 1cm ball at 4×

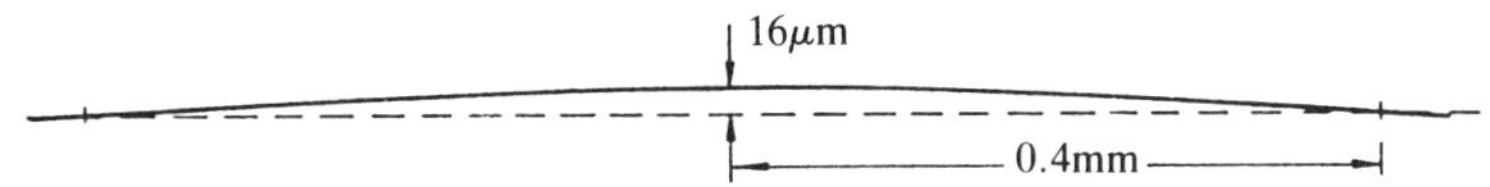

Fig. 1c 1cm ball at 200×

Fig. 1.

bed, but not be very good for a 1 cm ball — and you're wrong. If I draw the ball 4× life size it looks like Fig. 1b; but if I draw it, or part of it, at 200× we get Fig. 1c. If you're old enough to know the theorem of intersecting chords we can say at once that $b.b = h(2R-h)$, and if $h \ll 2R$ then $b^2 = 2RH$: so a rise of 16 μm takes place over a distance b = 0.4 mm, confirming that 1c is an honest picture. Still, that tiny departure from a plane *does* matter, so our second (first serious?) model is a parabola for a roller or a paraboloid of revolution for a ball: as far as the bits that could rub are concerned, a circle *is* a parabola.

The first introduction of this into tribology was by a certain H. Hertz, who later contributed to electrical theory; and he not only argued that when a ball is pushed against a plane, the parts which have to be deformed have a parabolic shape: he also argued that in calculating the stress distribution needed to deform them, he could take his ball to be a plane. So perhaps the boring model is a serious one, and the paraboloid is model 2.

Incidentally, in the messy problems which engineers seen to favour, the relevant local geometry is given by:

$$z = \tfrac{1}{2}Ax^2 + Hxy + 1\tfrac{1}{2}By^2,$$

and to model that we need to get the two curvatures A and B and the twist H correct. But in contact problems, unlike CAD, we are usually not interested in the orientation, so we can rotate the axes until there is no twist and we merely have two principal curvatures κ_1, κ_2 to describe the shape. To a first approximation we find that only $\kappa_1 + \kappa_2 = A + B$ is involved in the Hertz theory for the contact pressure and the compliance; or one can gain a factor of two in accuracy by using $\sqrt{(\kappa_1\kappa_2)} = \sqrt{(AB - H^2)}$ and pretending the shape is $z = \tfrac{1}{2}\sqrt{(\kappa_1\kappa_2)}(x^2 + y^2)$ — that is, modelling the surface as a paraboloid of revolution.

If we write $1/R$ for either $\tfrac{1}{2}(\kappa_1 + \kappa_2)$ or $\sqrt{(\kappa_1\kappa_2)}$, the contact radius according to Hertz is given by $a = (3WR/4E^*)^{1/3}$ and the maximum pressure by $p_0 = 1/\pi(6WE^{*}/R^2)^{1/3}$ — the alternative values of R giving upper and lower bounds. For moderately elliptical contacts this is much more convenient than using tables of elliptic integrals to get exact answers. $1/E^* \equiv (1 - \nu_1^2)/E_1 + (1 - \nu_2^2)/E_2$ defines the contact elastic modulus of the two bodies: we shall meet it again.

Note that for $R = 10$ mm, steel on steel ($E^* = 115$ kN/mm^2), a load of 1 kN gives a maximum pressure of 3 kN/mm^2 (just allowable with hardened steels) and a contact radius of 0.4 mm, which is why we only need to know the shape immediately around the first point of contact.

But the real problem of surface modelling in tribology, again unlike CAD, is not how to model a curved surface: it is how to model a *plane* surface. I mean, of course, a real, *nominally* plane, surface. Suppose we look at a metal surface (Fig. 2). It

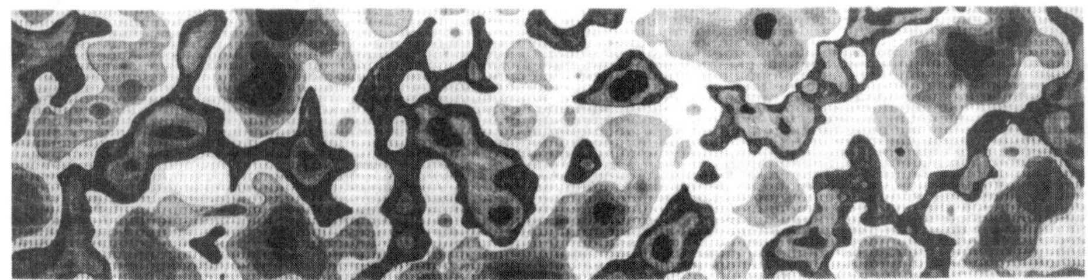

Fig. 2 — Map of bead-blasted aluminium surface.

resembles the surface of the Earth: it is covered with bumps, and if we try to rub two surfaces together, the bumps — asperities is the accepted term in tribology — will certainly play a part. In the words of my first Professor:

> Putting two solids together is rather like turning Switzerland upside down and standing it on Austria — the area of intimate contact will be small

Actually, he was wrong: Switzerland is too rough, and Southeast England is more typical.

It was probably Coulomb who introduced surface roughness into tribology, and he claimed that friction was caused by the inclined plane mechanism: as the asperities climb over each other, there is a component of the normal reaction opposing motion. He apparently never faced up to the question of descending the other side: but more to our point, he never said anything quantitative about the roughness. Indeed, Coulomb's identification of friction with roughness led to the misappropriation of the term 'rough', and so made the study of roughness more difficult. For if you want to publish a paper on the contact of a sphere with a rough plane, who will believe you mean a frictionless, bumpy plane? Explaining to an applied mechanics man that by 'rough' you mean 'not smooth' hardly helps: of course 'smooth' is synonymous with 'frictionless'. Not, however, in the real world.

The next model logically is to take your nominally plane surface to be wavy. Metrologists separate surface roughness into 'true' roughness and 'waviness', and it is in the nature of most metal-shaping processes that the surface height includes a distinct periodic component — rather obviously with a turned surface, usually detectable with a ground surface.

If a surface may be modelled as a plane with a single superimposed sinewave, it is relatively easy to analyze the contact. For example, Blok (1950) showed that a plane, sinusoidal surface $z = d\cos(2\pi x/\lambda)$ can be flattened by a sinusoidal pressure distribution $p = p^* \cos(2\pi x/\lambda)$ if $p^* = \pi E' d/\lambda$ where E' is the plane strain modulus, and that the maximum shear stress is p^*/e. A convenient measure of the strength of the surface is the hardness H, which is related to the plastic yield stress in shear k by $H \approx 5k$: so Blok concluded that the waviness can be flattened elastically provided that:

$$d/\lambda < \frac{e}{5\pi} \cdot \frac{H}{E'} \dagger.$$

If both surfaces are elastic, then replacing E' by E^* gives the condition for complete conformity by elastic deformation.

If we are trying to represent roughness by regular waves, it is probably more realistic to consider waves in two directions and write:

$$z = d\cos 3\pi x/\lambda + d\cos 2\pi y/\lambda . \dagger$$

† Blok actually suggested $(2/\pi \cdot)H/E'$ which is of course the same as: $z = 2d\cos(\pi\sqrt{2}/\lambda \cdot x + y/\sqrt{2})\cos(\pi\sqrt{2}/\lambda x - y/\sqrt{2})$, that is, the product of two waves of wavelength $\lambda\sqrt{2}$.

The condition for elastic conformity then becomes:

$$d/\lambda < \frac{1}{3.1\pi} \cdot H/E^*$$

Now the maximum slope of the sinusoidal ridges is $m = 2\pi d/\lambda$ (or $2\pi d\sqrt{2}/\lambda$ for the 2-D roughness), so that one can state the result as:

1–D max. slope $< 1.09\ H/E^*$

2–D max. slope $< 1.90\ H/E^*$

Thus, for mild steel with a hardness of 100 (1 kN/mm^2), the slopes must be less than half a degree to permit elastic conformity — in other words, almost any two surfaces placed in contact will deform plastically.

Blok's condition for elastic flattening, rewritten as $mE/H^* < 1$, is the first example of a 'plasticity index' in tribology. If the plasticity index is low, we get elastic contact: if high, plastic. Plastic contact certainly has one advantage: it permits us to calculate the real area of contact by simply dividing the load by the hardness: we may forget all about surface modelling. But we believe that plastic deformation normally fractures the brittle oxide, permitting real metal–metal contact, which leads to adhesion, high friction, metal transfer — in short, to a risk of failure. For safety, we need elastic deformation, which the oxide film can normally accommodate without fracturing, and which leaves the surface as it was — the surfaces will last for ever, or at least until the fatigue limit is reached.

Blok's condition is too stringent. We do not actually require the surface roughness to be levelled to permit sliding without damage to occur: we merely require that any deformation which does occur should be elastic. This is obviously true for lubricated sliding, in which we have an oil film largely separating the two surfaces and only occasional asperity contacts, but it is also the case with dry sliding. We shall find a more suitable condition later.

This model of a doubly-wavy surface may well be the way to begin the study of a neglected topic in tribologhy: 'lubrication by puddles'. You will not find this in the books: but consider what happens when such a surface is loaded against a plane (Fig. 3). The initial contacts at the summits grow and join up to form a seal round the valleys [Johnson *et al.* (1985)]. If the surfaces are coverd with lubricating oil, oil will be trapped in the valleys, and will develop high pressure and so make a major contribution to supporting the load, while the puddles also act as reservoirs which provide 'weeping lubrication' over the nominally dry contact areas.

Now we must improve our model of a surface. Better models — or at least, more complicated ones — did exist before Blok's contribution, mostly in the Soviet literature, and postulated that a rough surface is a plane surface with hills on it. No serious measurements of surface roughness existed to cloud the issue — the surface profilometer was invented only in 1932 — became a commercial instrument only in

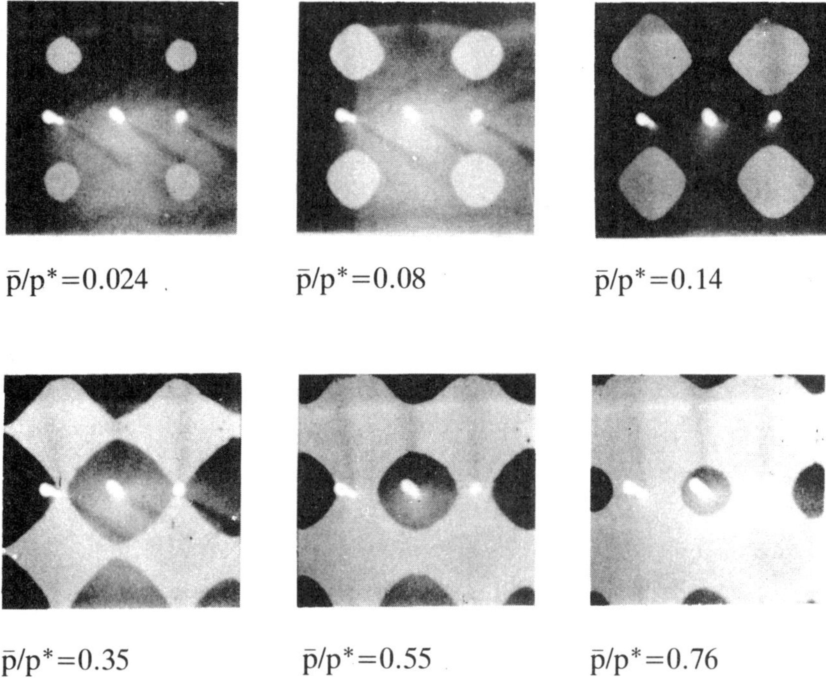

Fig. 3 — Experimental contact areas (appearing light).

1937 — and was used only as a quality control device until the late 1950s. So let's choose a convenient shape of hill — a spherical cap of given radius of curvature — and scatter some on our plane. The essential variability is introduced by assuming the heights to differ, though only in the simplest way: rectangular distribution (equally probable heights over some interval $0<z<H$) or a triangular distribution (probability linearly decreasing to zero over $0<z<H$) (Figs 4a, b). Incidentally, if we have two surfaces with triangular height distributions, and we assume the asperities are aligned (but why should they be?), contact will depend on the joint height distribution (Fig. 4c) — we can see the Central Limit Theorem of statistics busily making the sum of independent random variables converge toward a normal distribution — and in particular, producing the long tail which turns out to be vital.

Zhuravlev (1940), to whom this analysis is due, showed that if each individual contact deforms according to the Hertz laws (area proportional to two-thirds power of the load), the total area increases almost in proportion to the load (actually as $(\text{load})^{10/11}$) that is, the basic Amontons' law of friction, which is believed to depend on the real area of contact being proportional to the load, does not rely on the area of contact being determined by plastic deformation. The second question, of whether the deformation will be elastic or plastic, Zhuravlev did not consider.

Ling (1958) put forward a similar model; a rectangular height distribution, but using cones with semi-angle 45° for his hills. This was really rather unjustifiable in

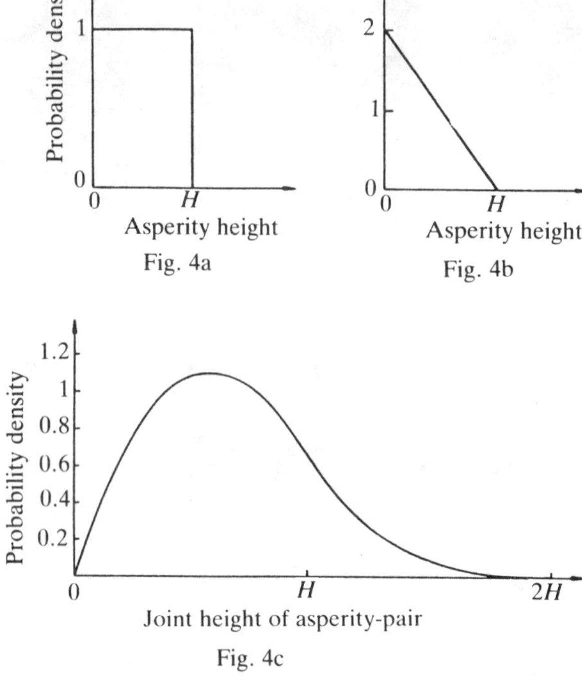

Fig. 4 — Simplistic height distributions.

1958, when profilometry had certainly established that most slopes were very much less than this; but Ling must certainly be given the credit for realizing that the load–compliance relation for the surfaces could be predicted theoretically and compared with laboratory mesurements — one of the few methods available for actually studying surface contact because of the difficulty of 'getting into the contact' to actually see it.

Though actually, you can get in, as shown by Dyson & Hirst (1954). If one of the contacting bodies is a glass block with a silvered surface, you can look at the underside of the silver film and see the deformation — provided that you have a microscope which is sensitive to small changes in slope. The contact between a smooth ball and a smooth aluminium film on a smooth microscope slide looks like this (Fig. 5) — and incidentally, the contact is perfectly elastic — or at least, when you roll the ball away, no traces are left.

By the early 1960s a number of groups saw that a more quantitative analysis of surface roughness was required, and connected their profilometers to a digital voltmeter, data-logger, and computer to start acquiring numbers on the grand scale. I think it was then that the basic statistical nature of surface roughness really came over; certainly up to then roughness standards based on the 'highest-peak' and the 'lowest valley' — of a profile containing perhaps twenty of each — existed, which is appropriate for a turned surface where the randomness may be a small imperfection

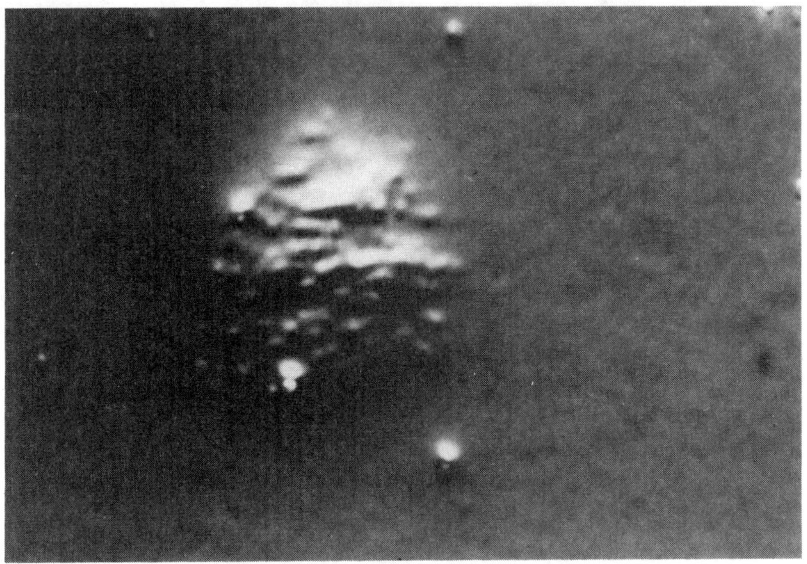

Fig. 5 — Contact between 'smooth' ball and 'smooth' aluminium film.

in the waviness, but inappropriate for most others. There seems to have been no recognition that the 'root-mean-square' height was a standard deviation of a height distribution, or that the 'centre-line-average' roughness was a 'mean deviation'.

It became clear that surface heights are distributed roughly 'normally' — or I prefer to say are roughly 'Gaussian' — but are probably not actually Gaussian (Fig. 6). So we hastily improve on Zhuravlev by assuming a Gaussian height distribution (still of identical spherical caps), and find that the area of elastic contact is now very nearly proportional to the load, and that we can get a new plasticity index:

$$\Psi = \frac{E^*}{H}\sqrt{\frac{\sigma}{R}}$$

where σ is the standard deviation of the summit heights and R the radius of curvature of the asperities (Greenwood & Williamson 1966). Note that with an unbounded height distribution we must always get some plastic deformation: it turns out that with this model the amount — the fraction of the area of contact which is at the yield point — remains almost constant as the load increases: $\Psi = 1$, is, more-or-less, the condition for 1% of the area to have yielded.†

It is worth spelling out how this can be, for as the load increases, the area of each contact, and the pressure on that contact, increase until that contact yields. But at the same time new, elastic, contacts are formed: at such a rate that the typical contact remains elastic, and indeed, such that its size remains constant.

Unfortunately, when we measure the asperity curvatures, the concept of a typical asperity quickly disappears (Fig. 7): the distribution of peak curvatures is close to a

† For $\Psi = 0.9$, there are hardly any plastic areas: For $\Psi = 1.1$, the proportion is already much more than 1%.

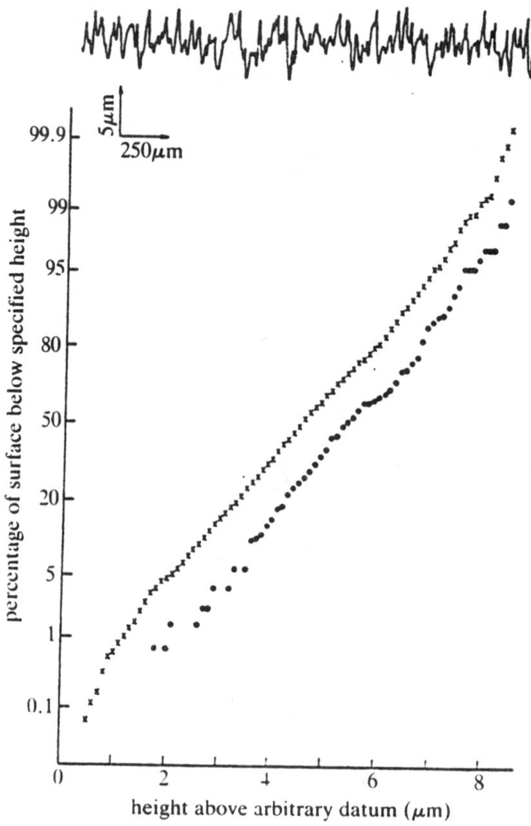

Fig. 6 — Cumulative height distribution of bead-blasted aluminium. Both the distributions of all heights (×) and of peak heights (•) are Gaussian, at least in the range ±2 standard deviations. The profile of the same surface is shown in the upper diagram: the vertical magnification is 50 times the horizontal magnification.

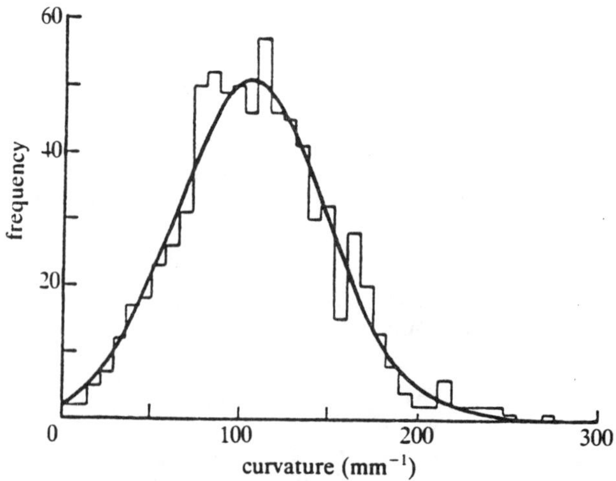

Fig. 7 — Histogram of curvatures at peaks for bead-blasted gold specimen. The continuous curve is the Gaussian with the same mean and standard deviation.

Rayleigh distribution — with a standard deviation equal to half (actually 0.523) of the mean, so that the mean is a poor representative. Note the importance of studying curvature, rather than its reciprocal; the radius of curvature: if curvatures follow Rayleigh (that is, $te^{-1/2t^2}$) then the product of the mean radius with the mean curvature is not 1 but $\pi/2$ — and the S.D. of the radius is infinite: the mean radius is useless as a representative value. This is bad enough: but the mean curvature — or mean radius — also varies with peak height: the higher peaks are more curved than the lower ones. And that really knocks out direct experimental measurements, for one needs to study what in profilometer terms is an impossibly large area of the surface before one can acquire an experimental curvature distribution for the high peaks.

I believe it was Tallian and his colleagues at SKF (1964) who first noted the connection between surface profiles and Rice's theory (1945) of noise in telephone lines — random signal theory: they observed that not only was the height distribution (on an SKF ball) Gaussian, but that the distribution of crossings of a given level followed the same Gaussian. Rice's theory predicts this — but the elementary interpretation is startling: the mean slope is the same at all heights — there is no levelling off as one reaches the top. Can the asperities be cones after all?

Surfaces are of course two-dimensional: only the one-dimensional profilometer trace is a random signal, and the random process theory we need is not that due to Rice but that produced by Longuett-Higgins (1957) to describe the surface of the ocean. Instantaneously it appears that the surfaces of a block of steel and of the ocean are much the same: we have the advantage that our surface stays still.

The random process model is that the surface height is the superposition of a continuous spectrum of waves with different frequencies, directions, and phases. It follows immediately that the height distribution is Gaussian: that the peaks encountered along a line are roughly but not quite Gaussian, while the summits — the features that actually take part in contact — are again roughly but not quite Gaussian, with, obviously, a slightly higher mean than the peaks and, less obviously to me, a slightly lower standard deviation. We have to thank Nayak (1973) for emphasizing that a peak is not a section through a summit, and not a particularly good representative of a summit: a typical traverse of a surface by a stylus will encounter a large number of shoulders but very few summits (Fig. 8).

Nayak (1971) added to Longuett-Higgins' analysis the feature the tribologist needs: the derivation of peak and summit curvature distributions. According to random process theory, peak curvatures should indeed follow a Rayleigh distribution: summit curvatures have fewer very low values and are slightly more concentrated round the mean values ($\sigma_\kappa \equiv 0.4\,\mu_\kappa$ for summits compared to $\boldsymbol{\sigma}_\kappa \equiv 0.5\,\mu_\kappa$ for peaks) — but vary almost as strongly with height.

The really astonishing thing about random process theory is that all the properties of the surface are related: three parameters, defined by Longuett-Higgins and Nayak as the moments of the power spectrum of the waves which form the surface, are sufficient to describe all the properties of the surface needed to study contact. I shall use a more accessible definition and take the parameters to be the rms height σ, the rms slope of the profile, σ_m, and the rms curvature of the profile, σ_κ: in fact we have $m_0 = \sigma^2$, $m_2 = \sigma_m^2$, $m_4 = \sigma_\kappa^2$. σ merely plays the role of a scale factor, so the *nature* of the surface roughness depends on only two parameters, and many features depend

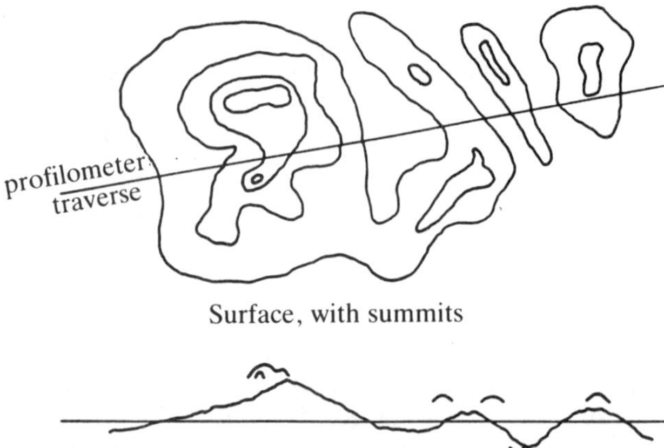

Fig. 8 — Relation between peaks and summits.

on only the single quantity $\alpha = (\sigma\sigma_\kappa/\sigma_m^2)^2$. For example, Nayak shows that the peak and summit height distributions, scaled by dividing by σ, depend only on α. They remain close to Gaussian, but, for example, the location and spread of the summit heights vary according to:

$$\bar{z}_s/\sigma = 4/\sqrt{\pi\alpha}; \quad \sigma_s^2/\sigma^2 = 1 - 0.897/\alpha;$$

that is, for large values of α there is little difference between surface heights and summit heights while for α approaching the minimum possible value ($\frac{3}{2}$) the difference is considerable.

Notice how far we have come from the idea of placing asperities on a plane — the same waves form the summits, the middle ground, and the valleys. It follows that the valley heights have the same distribution (except for a minus sign) as the summit heights, and that the valley and summit curvatures are the same — and in practice we find this holds for peaks and troughs on surface. All very satisfactory, it appears. But just in case you should think you understand, however dimly, what a surface looks like, I must introduce the work of Whitehouse & Archard (1969). They too model a surface as a random process, but as one with a specific autocorrelation function: they postulate that the correlation between heights falls off exponentially with separation — and they and others have shown that this is often an excellent approximation. The correlation function is the Fourier transform of the power spectrum; so this has implications for the moments $m_2(\equiv\sigma_m^2)$ and $m_4(\equiv\sigma_\kappa^2)$ — both are infinite. So is almost everything else — the mean peak curvature, the peak and summit densities, the zero-crossing density — and so the quantities one measures, necessarily at a finite sampling interval, are strongly dependent on that interval. Fig. 9 shows their measurements of peak curvature distributions: shall we take the mean peak curvature to be 4 mm^{-1} (sampling interval $h = 15\ \mu$m) or 250 mm^{-1} (h = 1 μm)?

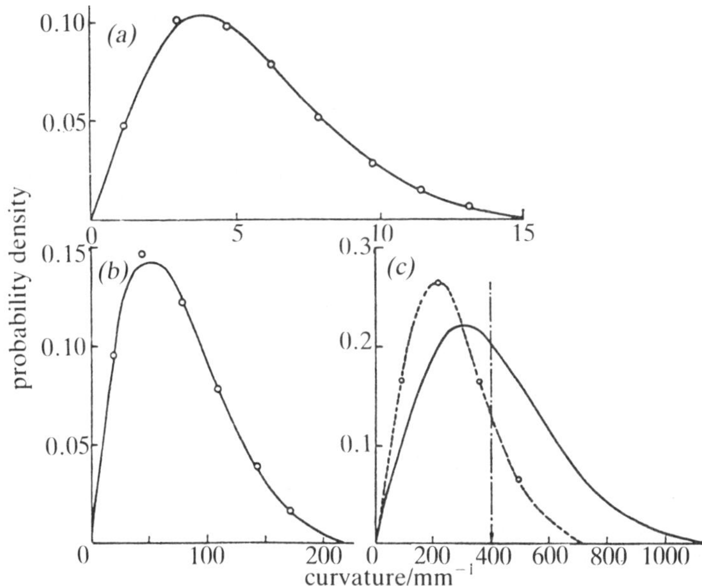

Fig. 9 — Probability densities of an ordinate being a peak of a given curvature. The full lines give the theory. The experimental points are from digital analysis of profiles from Aachen 64–13 ($\sigma = 0.5\,\mu$m, $\beta^* = 6.5\,\mu$m). Results are shown for three values of the sampling interval (l) corresponding to differing values of the correlation (ρ) between successive samples. (*a*) $l = 15\,\mu$m, $\rho = 0.10$; (*b*) $l = 3.0\,\mu$m, $\rho = 0.63$; (*c*) $l = 1.0\,\mu$m, $\rho = 0.86$; the arrow indicates the nominal stylus curvature, $400\,\mu\text{m}^{-1}$.

Onions & Archard (1973) studied the elastic contact between a surface of this nature and a plane, by rather arbitrarily choosing the properties measured at a particular sampling interval, that at which the readings were 'just independent' — actually where the correlation coefficient fell to 0.1. They also inadvertently used the properties of peaks rather than summits, so it is their qualitative results which are of greatest interest. They found that with this model, the already good proportionality between area of contact and load is improved, and that again the proportion of the area of contact which is plastic is almost independent of load, and governed by another plasticity index:

$$\Psi_A = \frac{E^*}{H} \cdot \frac{\sigma}{\beta}$$

where the correlation function is $C = \exp(-x/\beta)$; once again $\Psi_A < 1$ for largely elastic contact. This is a far more practical criterion than the earlier one: it may not be easy to select an effective β when the correlation function is only approximately exponential, and values of correlation around 0.1 are masked by sampling errors (see Hirst & Hollander 1974), but it is preferable to trying to select and justify a single characteristic radius of curvature for the Greenwood & Williamson plasticity index.

But what can one make of a surface where all the properties alter when we alter the sampling interval?

From the mathematical point of view, the situation is not too bad.

Whitehouse & Phillips (1978, 1982) and Greenwood (1984) have shown that a random process theory *at a finite sampling interval* can be developed which predicts all the quantities studied by Nayak, and that if quantities are suitably normalized, the results at a finite sampling interval are not very different from those at zero sampling interval. For example, the mean peak curvature at a finite sampling interval h is given by:

$$\bar{\kappa}_p/\sigma_\kappa = \sqrt{\pi/2}\cdot\sin\theta/\theta \tag{4}$$

where $\sin\theta = h\sigma_\kappa/2\sigma_m$. Rice's result corresponds to $\theta = 0(\bar{\kappa}_p = \sqrt{\pi/2}\cdot\sigma_\kappa)$, while for the worst possible case of $\theta = \pi/2$ we get $\bar{\kappa}_p = \sqrt{2/\pi}\,\sigma_\kappa$: for a realistic value $\theta = \pi/4$ the reduction is only a factor 0.9. In all cases the distribution about the mean is almost the same (Fig. 10).These results hold whatever the correlation function: if it is such that $\sigma_\kappa \to \infty$ as $h \to 0$, equation (4) still holds: $\bar{\kappa}_p$ tends to infinity also.

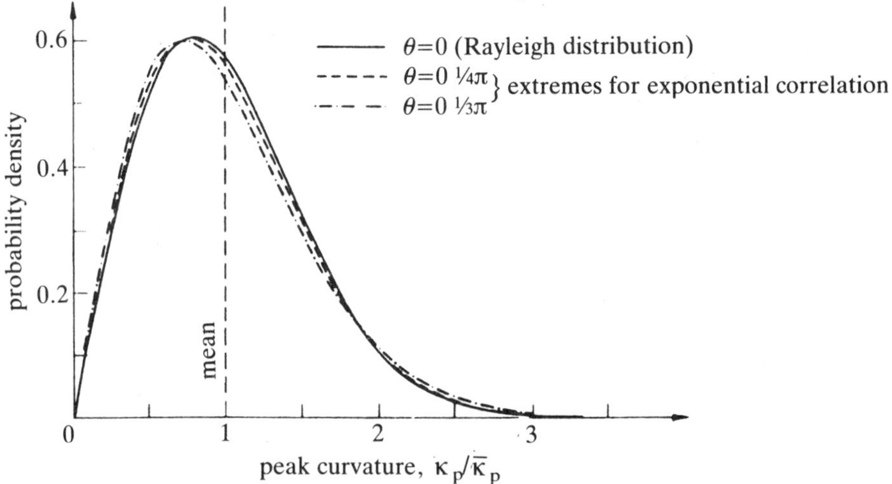

Fig. 10 — Peak curvature distributions for varying θ:$\theta = 0$ is the Rayleigh distribution $P(t) = te^{-1/2t^2}$; but note that the variable plotted is not $t = \kappa_p/\sigma_\kappa$ but $(\kappa_p/\bar{\kappa}_p)$.

Can we really have an infinite peak density and infinitely curved peaks? Surely not: we cannot meaningfully reduce the sampling interval below an atomic spacing (3×10^{-4} μm). But where we currently stop (1 μm) we have little comfort (Fig. 11): there is no satisfaction in knowing that the mean peak curvature lies between 1.25 σ_κ and 1.13 σ_κ if we cannot decide whether σ_κ is 100 mm^{-1} or 2 mm^{-1}.

Mandelbrot's exposition of fractal theory (1977), and his view that fractal distributions are the way of nature, whether we think of the perimeter of a snowflake

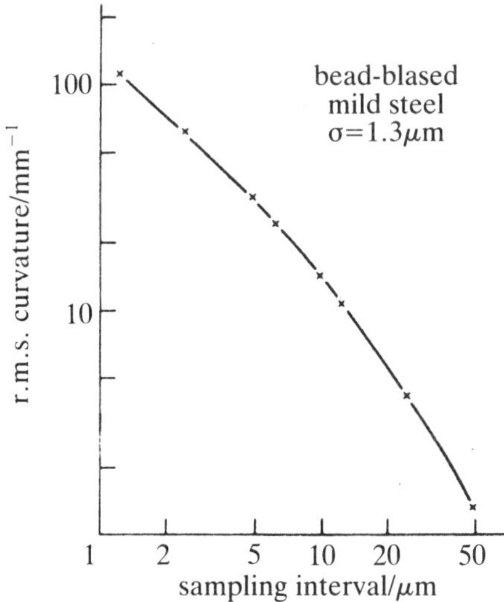

Fig. 11 — Variation of σ_K with sampling interval. Note that when a real property is considered, a logarithmic scale is desirable.

or of the British Isles, has done something to reconcile me to what we have found about surfaces: for the infinite densities, slopes, and curvatures we find are equivalent to saying that however much you increase the magnification with which you examine a surface, what you see doesn't change: the fractal principle.

Sayles & Thomas (1978) have an alternative line of justification: a truly exponential correlation function corresponds to a power spectrum $G(k) = C\lambda^2$ where $k = 2\pi/\lambda$. It is rather difficult to measure the amplitude of a 1-cm wave with a profilometer — especially on a specimen 2 cm long: but let's assemble all the information on the roughness of different bodies measured, using different instruments (Fig. 12). The Moon's surface, and a surface ground specimen, all lie on the same curve: $G(k) = C/\lambda^2$ is a law of nature. So the question is, not how do we model the surface, but how do we model the *response* to the surface?

It appears, then, that an actual surface is a horrible thing, and only if we observe it through suitable blinkers will it be all right. Of course all physical interactions involve some form of blinker: for example the profilometer itself more-or-less averages the surface height over the area of the stylus tip: so the power spectrum received is no longer $G(k) = 4\pi^2 C/k^2$ but $F(k) = 4\pi^2 C/k^2)\,(\sin kb/kb)^2$ where $2b$ is the length of the stylus tip — and the correlation function stops being exponential below b: all our problems have disappeared. This, then, is the way forward: consideration of the choice of blinker. So far we haven't made much progress — but in any case that's no longer surface modelling.

ACKNOWLEDGEMENTS

Fig. 2 was kindly supplied by Dr J. B. P. Williamson; Fig. 3 is reproduced by permission of the *Iternational Journal of Mechanical Sciences*; Figs 6, 7, 9, 10, 11 by permission of the Royal Society, and Fig. 12 by permission of *Nature*.

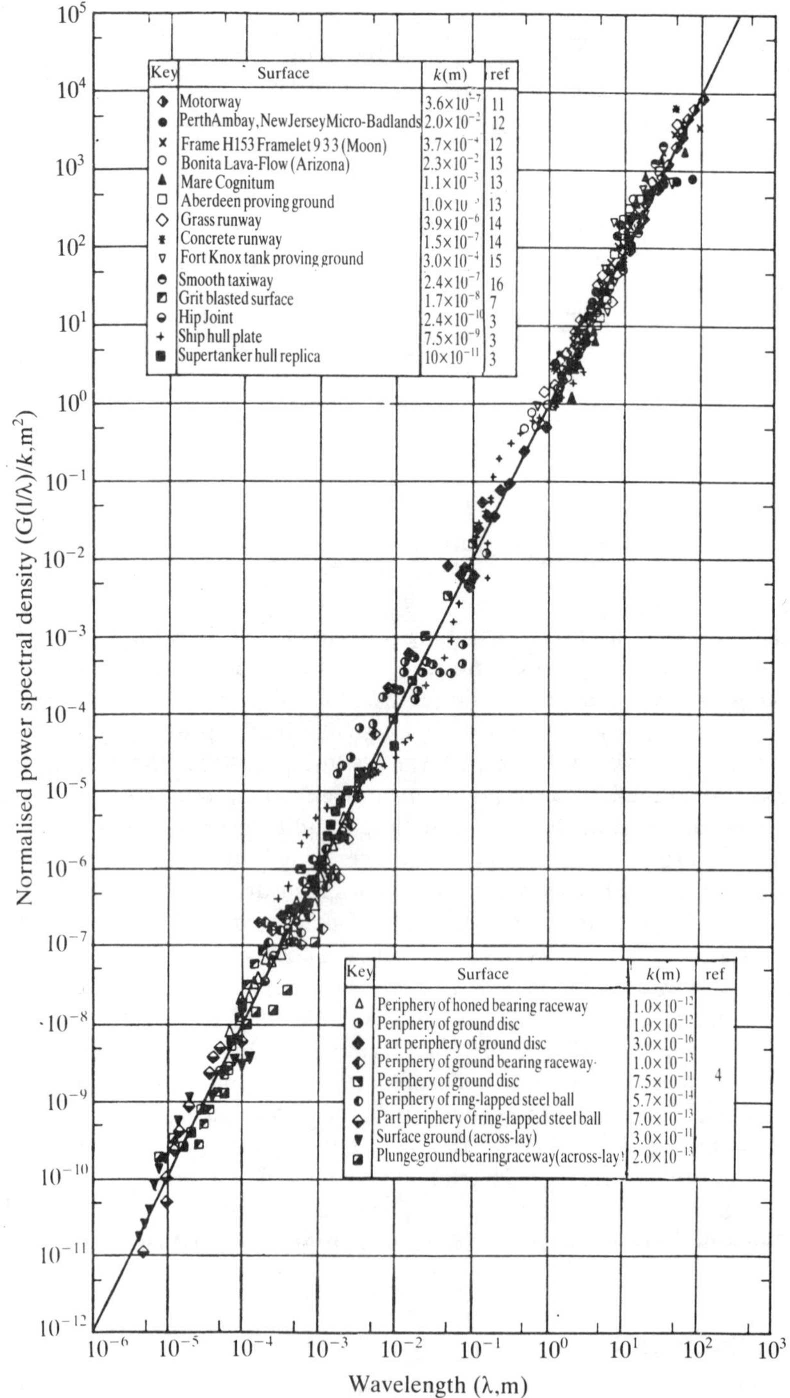

Fig. 12 — Spectral density of 'everything'.

REFERENCES

Blok, H. (1950) Discussion on friction. *Proc. Roy. Soc. Lond. A* **212** 420.

Dyson, J. & Hirst, W. (1954) *Proc. Phys. Soc.* **67 B** 309.

Greenwood, J. A. & Williamson, J. B. P. (1966) *Proc. Roy. Soc. Lond. A* **295** 300.

Greenwood, J. A. (1984) *Proc. Roy. Soc. Lond. A* **393** 133.

Hirst, W. & Hollander, A. E. (1974) *Proc. Roy. Soc. Lond. A* **337** 379.

Johnson, K. L., Greenwood, J. A. & Higginson, J. G. (1985) *I. J. Mech. Sci.* **27** 403.

Ling, F. F. (1958) *Trans. A.S.M.E.* **80** 1113.

Longuett-Higgins, M. (1957) *Phil. Trans. Roy. Soc. A* **249** 321 and A **250** 157.

Mandelbrot, B. (1977) Fractals: Form, chance and dimension, Freeman, N.Y.

Nayak, P. R. (1971) *J. Lub. Tech.* (*ASME*) **93F** 398.

Nayak, P. R. (1973) *Wear* **26** 165.

Onions, R. A. & Archard, J. F. (1973) *J. Phys. D. Appl. Phys.* **6** 289.

Rice, S. O. (1944) *Bell System Technical Journal* **23** 282 and **24** 46 (1945).

Sayles, R. S. & Thomas, T. R. (1978) *Nature* **271** 431.

Sayles, R. S. & Thomas, T. R. (1979) *J. Lub. Tech.* (*ASME*) **101F** 409.

Tallian, T. E., Chiu, T. P., Huttenlocher, D. F., Kamenshine, J. A., Sibley, L. B. & Sindlinger, N. E. (1964) *ASLE Trans.* **7** 109.

Whitehouse, D. J. & Archard, J. F. (1969) *Proc. Roy. Soc. Lond. A* **316** 97.

Whitehouse, D. J. & Phillips, M. J. (1978) *Phil. Roy. Soc. Lond. A* **290** 267.

Whitehouse, D. J. & Phillips, M. J. (1982) *Phil. Roy. Soc. Lond. A* **305** 441.

Zhuravlev, V. A. (1940) *J. Tech. Phys.* (*USSR*) 10 1447.

8

Modelling dental surfaces

M. Bramwell
Thames Polytechnic, School of Maths, Stats and Computing, UK
R. DeLong, R. Sakaguchi, M. Pintado and **W. Douglas**
University of Minnesota, USA

INTRODUCTION

The assessment of dental restorations entails both clinical observations and analysis made in the laboratory. We need to be able to predict the behaviour of dental materials, both natural (enamel dentine) and artificial (amalgam, composite resins, and porcelain). This may require data from clinical observations, laboratory analysis of replicas of clinical restorations, laboratory simulations of clinical conditions, and possibly even theoretical analysis.

IN THE CLINIC

The dentist working in the clinic has three main sources of information.

(1) Visual inspection of the surface of the tooth.

Such information is processed very rapidly by the brain, but there are limits to what the eye can perceive. Gross discrepancies such as a fractured cusp of a restoration are readily seen, but subtle changes of the surface may not be detectable with the unaided eye.

(2) Information sampled at a point, or rather a small region of the surface, using a probe.

In this case a defect in the surface of the tooth can be confirmed by tactile sensation.

(3) A projection of the tooth and its interior can be obtained by radiographic techniques.

It is difficult to extrapolate from a single two-dimensional image to obtain information about a three-dimensional solid or a surface. However, the image will show projections of the boundaries of regions within the tooth and will reveal, in particular, the presence of dental caries and defects in the margins of restorations.

A few comments can be made about the quality of this information.

(1) It is qualitative rather than quantitative.
(2) It refers to past events, and it does not allow us to predict the future behaviour of the tooth or material, other than the necessity of impending replacement or repair.
(3) Useful general information of a statistical nature can be deduced, such as that material A tends to chip, or that restorations using material B seem to have an adverse effect on restorations using material C.

IN THE LABORATORY

Several pieces of equipment have been developed in the Biomaterials Program of the University of Minnesota to facilitate the evaluation of dental restoratives (Fig. 1).

(1) Simulation of chewing (DeLong & Douglas 1983)

Two types of 'artificial oral environments' were developed to reproduce the forces, movements, and environmental conditions of the human chewing cycle. The equipment can be used to abrade natural, freshly extracted teeth in a fashion that simulates the chewing action of a human mouth. This even includes the provision of artificial or natural saliva at body temperature. These 'artificial oral environments' can chew continuously twenty-four hours a day, seven days a week, so that the production of wear data under simulated physiological conditions can be greatly accelerated. One year's wear of a restoration in a human mouth can be duplicated in the 'artificial environment' in less than twenty-four hours! By placing strain gauges on the tooth it is possible to obtain information about the strain and hence the stresses in the tooth under various physiological loading conditions.

(2) Three-dimensional surface digitization (Delong *et al.* 1985(a))

A dental restorative material which loses more than 50 μm of height in a single year in the mouth is deemed unacceptable. Thus, producing wear is only part of the problem; the second part is to be able to measure accurately the amount of material removed. To accomplish this a three-dimensional digitizing system was developed which is capable of digitizing the highly irregular anatomy of the surfaces of teeth. Measurements are made by physical contact with a stylus. The system can output (x,y,z) coordinates of equally spaced points along profiles on the surface of the tooth with a precision of better than 7 μm (De Long *et al.* 1988). Precision, in this context, means the accuracy with which a profile can be remeasured. The stepper motors controlling the horizontal motion are accurate to within 1 μm. On a flattish portion of the surface the horizontal discrepancy should not be more than 0.5 μm, but on a steep incline this can increase to about

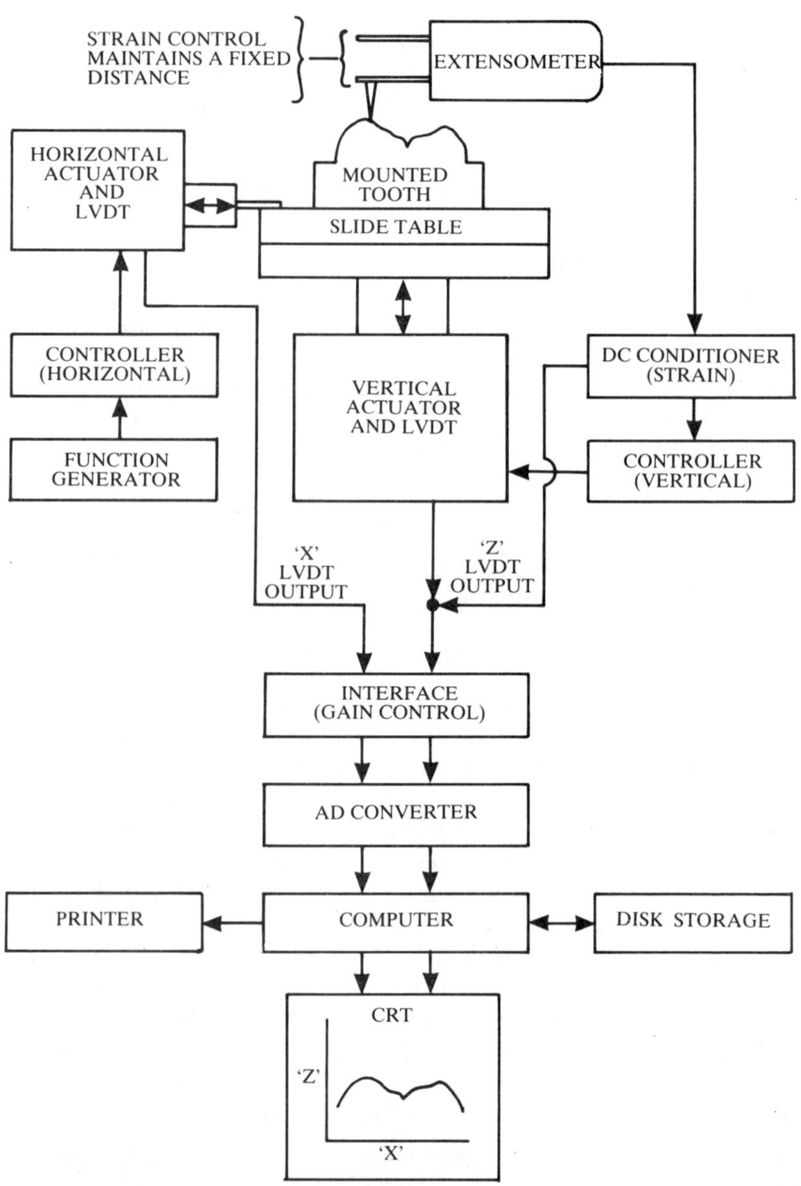

Fig. 1 — Block diagram of the profiling equipment reprinted from DeLong *et al.* (1985). The tooth, mounted on the slide table, maintains contact with the fixed stylus.

7 μm. The surface information is then available as points on a mesh over a regular grid. The equipment is used to evaluate dental materials, both natural and artificial, in a simulated environment in order to gain an understanding of how they function, and, if possible, how they can be improved. It has also been

used to evaluate dental materials clinically by measuring surface changes on very accurate casts made from impressions of the dental materials. Both of these uses have proven very beneficial in reducing the time and costs of clinical studies for the evaluation of different dental restorative materials.

ORIENTATION AND OTHER MEASUREMENT PROBLEMS

The normal surface anatomy of a tooth is highly irregular, having sharp cusps, steep inclinations, and many deep pits and grooves. This makes the surface very difficult to digitize accurately and rapidly. Although the contact stylus method seems to provide a good solution, it is not, however, without problems. Contact with the surface can cause surface distortion which can also lead to surface damage and inaccurate measurements. Fortunately, most of the surfaces of concern in dentistry are very stiff. A second problem is the shape of the stylus.

Where the surface is steep, for example near the incline of a cusp of a molar or near the sides of the tooth, a small lateral displacement (dx, dy) will produce a large vertical displacement dz. Thus errors can arise from the finite size (radius) of the stylus tip. Further, because the stylus is in contact with the surface it can wear. Thus its shape, which is assumed to be spherical, may not in fact be so. This last source of error can be virtually eliminated by use of an 'optical stylus', which does not contact the surface. Such a stylus is under development. This will also reduce the other sources of error since they are clearly related.

Where the equipment is used to monitor wear by using casts made from impressions of a tooth in a human mouth at, say, intervals of six months, a more serious problem arises. To make comparisons the casts must be oriented in exactly the same way. When simulation studies are done in the laboratory by using the 'artificial oral environment', orientation is not a problem, because the mounts holding the teeth have precision guides which allow very accurate alignment of the teeth. These precision guides do not exist in the human mouth. There are two possible approaches to orienting the models. Either the teeth could be modified in the mouth in some way to provide points for alignment, or the very irregularity of the surface anatomy of the tooth can be used as a means of alignment. From our point of view the former option is not an option. Techniques were therefore developed to use the surface anatomy of the teeth to align the casts prior to digitization. Unfortunately this mounting procedure is not perfect, and misalignments of up to 100 μm can still occur. This can lead to significant errors in measurement.

To reach the desired accuracy of 10 μm or better, the orientation error must be eliminated. In fact the success or failure of the entire measurement system depends on how well this reorientation can be achieved, and not on how accurately the system can digitize a surface. Precise manual alignment of the surfaces would require much time and effort, whether this was attempted by redigitizing or by manipulating the displayed images of the surfaces. In addition to creating software for data collection and display, it has thus also been necessary to write software to align the surfaces, using only the anatomy of the teeth. At this point the irregularity of the surfaces of the teeth again becomes very beneficial. The software for display and alignment, like the equipment, has gone through several generations, migrating from the Apple II to a Masscomp minicomputer to an AT and finally to a 386. Currently, the data can be displayed as a surface with or without hidden line removal, as individual profiles in the (x,z) or (y,z) planes, as a contour map or as a relief map. In addition to the surfaces to be aligned, the difference surface can also be displayed. The various displays are used to enable the operator to input information as is necessary to facilitate the computer alignment of the surfaces.

ESTIMATING SURFACE CHANGE

Because two similar surfaces cannot always be aligned perfectly before digitizing, techniques for estimating change in the surface contour must include the following:

(1) The transformations required to fit the surfaces must be determined.
(2) One of the surfaces must be moved so that the differences between the surfaces can be determined.
(3) As the surfaces are 'known' only at discrete points over a grid whose points are 50 to 100 μm apart, some form of interpolation is required. In deciding which type of interpolation to implement, it is necessary to take into account the conflicting requirements of accuracy and speed.
(4) An accuracy of better than 10 μm is necessary.

Further, all techniques must take account of the fact that points on the surfaces fall into three classes:

(a) Points representing unaltered parts of the surfaces. These are used to determine the transformation for the alignment process.
(b) Points alterations are likely to have occurred. These will include wear facets on a tooth and broken restorations, but may also include bubbles due to imperfections in taking impressions or making casts.
(c) Points where measurements are suspected to be inaccurate, such as points measured on steep inclines. These should be ignored. The processes required for alignment can be applied either iteratively in sequence or as once and for all operations. Once the surfaces have been fitted, measurement of volume loss is straightforward. In addition to a numerical evaluation of volume loss, the software allows the loss of depth to be displayed as a relief map on the anatomical surface of the tooth. To date, software has been developed based on three distinct principles. In all cases one surface is regarded as being fixed and the other surface is then fitted to it.

(1) Profiling
Each profile is fitted to the corresponding profile on the fixed surface in turn. Here cubic splines over the fitting regions form a natural method of interpolation. The matching process uses an RMS minimization of the distances between appropriate points on the curves. The procedure is iterative for each profile, with translations and rotations being applied in the plane of the profile. The technique is slow by its very nature as there may be some hundred profile pairs, but it can be expected to produce satisfactory results, provided that large rotations about a vertical axis have not occurred. It is clear that as the operations take place in a fixed vertical plane, rotations about the z-axis cannot be catered for. As the two profiles can be displayed on the screen together, an experienced user can easily select regions to be used in the matching process, and it can be seen whether a fit

has been achieved by inspection of profiles in both the (x, z) and the (y, z) planes. Because the order of processing can keep step with the order of data collection, this method is well adapted to coping with dislocations in a measured surface caused by electronic shifts (Fig. 2).

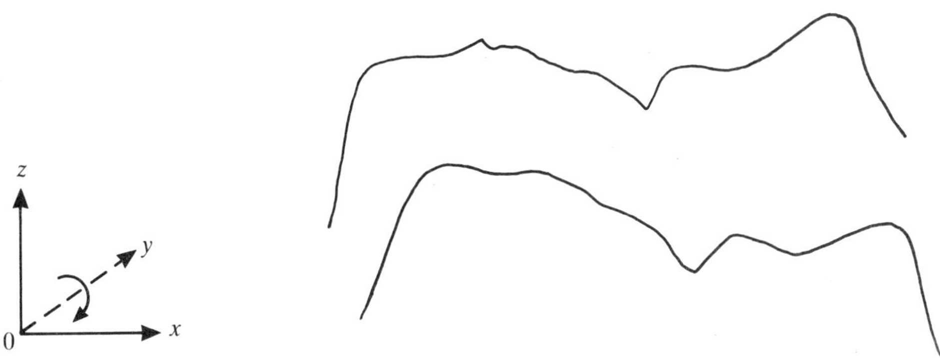

Fig. 2 — Method (i) is applied to each profile pair in turn. To fit the profile the required translation and rotation are obtained by summing adjustments x_i, z_i and θ_i under user control. The goodness of fit obtained can be assessed visually as well as numerically.

(2) Three point fitting (DeLong & Bramwell 1987)
Three points are selected on the fixed surface, and a match is established by scanning the second surface in the region of the corresponding points. Once a match has been found the transformations can be determined, so that fitting can be achieved as a once and for all process. The criteria used for matching essentially compare the flatness of regions near corresponding points. Hermite interpolation is used, the first and second derivatives being estimated from a simple approximation scheme. When a match has been achieved with this technique, a good fit has resulted. Unfortunately the procedure, in the present state of the software, seems to be very sensitive to small local variations in the fixed surface, so it is unreliable, as even skilled operators are unable to make a consistent selection of points. It should be possible to much improve the reliability, without loss of performance of the method, by using parallel processing techniques.

(3) Statistical surface fitting
A matching region is selected from a contour map of the differences over the whole of the surfaces. In general this region will be made up of disjoint sections. Regions where unusually large differences occur are avoided. An rms fit is then obtained, iteratively, for the whole of the region. Some skill is needed to select both the fitting and the wear regions, as the contour map is not as easy to read as an anatomical surface. In this case interpolation is currently using quadrilateral patches that are simply convex combination of the vertices. A good strategy is needed to control the iterations. The method used here is a species of steepest

descent based on the simplex algorithm (Caceci & Cacheris 1965, Nelder & Mead 1965). This technique has proved to be generally reliable in practice. If convergence of the iterations is unsatisfactory this would normally imply that the fitting regions had been badly chosen (Fig. 3).

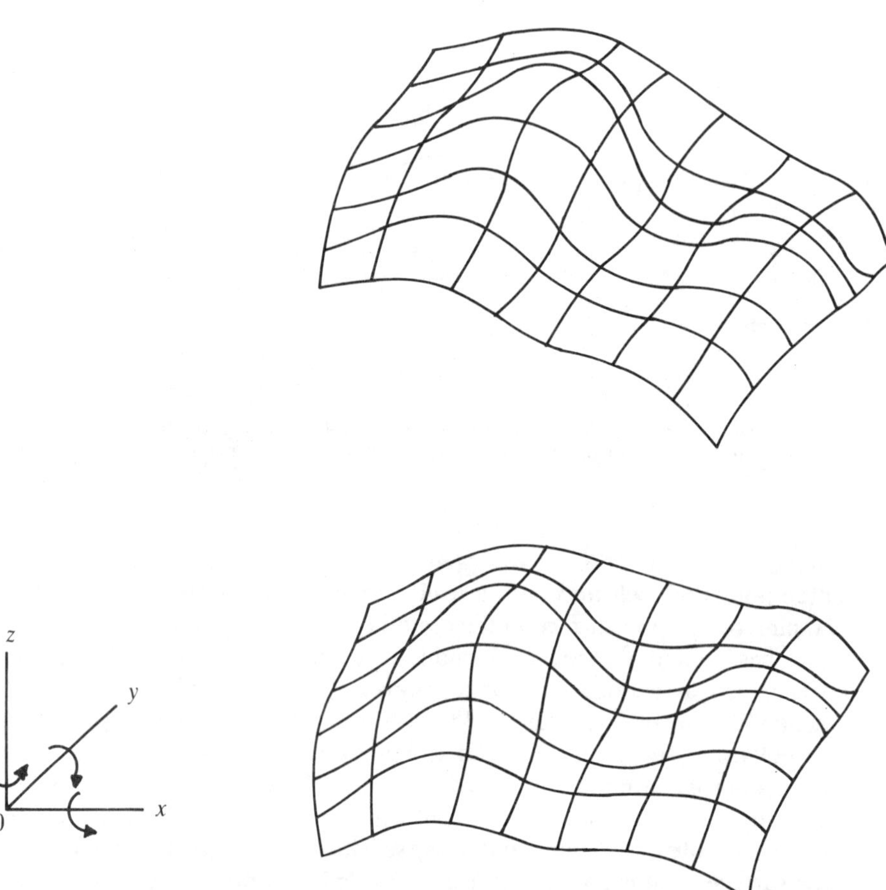

Fig. 3 — A full three-dimensional fit requires a translation (a, b, c) and three rotations of θ, ϕ, ψ about Ox, Oy, Oz respectively. Method (ii) attempts to find (a, b, c, θ, ϕ, ψ) by a single complex operation. Method (iii) improves the fit by a strategically chosen sequence of adjustments. The fit is tested at positions obtained by making increments in each variable in turn. Essentially an improvement is sought at a certain position in the opposite direction from the worst result. If this fails to improve the fit, then the increments are reduced.

APPLICATIONS

Many of the applications for which the system has been used to date have been concerned with properties of the dental materials used in restorative dentistry.

Currently, mercury amalgam is the most commonly used material in fillings. Porcelain crowns are commonly used for badly broken-down teeth, both for reasons of aesthetics and to restore the function of these teeth. Over the past few years composites, which consist of inorganic filler particles embedded in an organic resin, have been investigated. The properties of such materials depend not only on their chemical nature, but also on the geometrical distribution and size of the particles. Essentially, the closer the packing the greater is the compressive strength of the composite (Cross *et al.* 1983, 1985).

Extensive studies have been carried out on the wear of natural and artificial dental materials (Sakaguchi 1988, DeLong *et al.* 1985b, Sakaguchi *et al.* 1986, DeLong *et al.* 1986a). Cases where an artificial material is opposite natural enamel are of particular importance, since a badly designed restoration has the capacity to damage the opposing teeth. It has been shown that some composites are prone to wear, and that they characteristically exhibit 'dishing' in the vicinity of the region of contact with an opposing tooth. Some modern composites have better wear characteristics, but they exhibit fatigue problems which will limit the life of restorations. Experimental evidence also suggests that some composites are prone to chipping. Amalgam seems to resist occlusal wear quite well. In this case the primary wear mechanism is the transfer of material to the opposing cusp. Thus, after quite rapid initial volume loss the situation stabilizes (Douglas *et al.* 1985). Porcelain has been shown to be destructive to opposing enamel. The situation is exacerbated by bad occlusion that will result from a poorly fitted restoration.

To evaluate how a filling material supports the walls of a tooth under the stresses set up by chewing, strain gauges have been attached to extracted teeth. The measurements obtained have been used with a theoretical model using finite element methods. The results suggest that composites provide better support for the remaining tooth structure than amalgam (Douglas *et al.* 1985, Morin *et al.* (1988a). This occurs because composites can be 'cemented' to the remaining tooth structure, while amalgam cannot.

An ongoing series of studies is being undertaken which evaluates the performance of restorations in actual patients. The data are obtained from clinics as far as 1000 miles away. Casts are taken at periods from 6 months to 2 years and are sent to the laboratory for measurement and analysis (DeLong *et al.* 1986b).

The above by no means exhausts the applications of the Artificial Mouth and its software. Fissure sealants are regarded as beneficial in preventing dental caries, but are comparatively little used. There is currently some uncertainty about the retention and wear of such sealants. The techniques used above have been used to view the applied sealant and to calculate its volume. This provides a basis for the investigation of the changes of sealant with time (Pintado *et al.* 1988).

Finally, a truly exotic application: an Australian visitor to the Department has applied the equipment to the study of the wear of koala bear teeth!

REFERENCES

Caceci, M. S. & Cacheris, W. P. (1984) Fitting curves to data. *Byte*, May, 340–362.

Cross, M., Douglas, W. H. & Fields, R. P. (1983) The relationship between filler loading and particle size distribution in composite resin technology. *J. Dent. Res.* July, 850–852.

Cross, M., Douglas, W. H. & Fields, R. P. (1985) Optimal design methodology for composite materials with particulate fillers. *Powder Technology* **43** 27–36.

DeLong, R. & Bramwell, M. C. (1987) Matching an initial surface to a worn surface subject to rotation and translation. In: *The mathematics of surfaces II*, Oxford, 137–149.

DeLong, R. & Douglas, W. H. (1983) Development of an artificial oral environment for the testing of dental restoratives: bi-axial force and movement control. *J. Dent. Res.* **62** 32–36.

DeLong, R., Pintado, M. R. & Douglas, W. H. (1985) The measurement of change in contour by computer graphics. *Dent. Mater.* **1** 27–30.

DeLong, R., Sakaguchi, R. L., Douglas, W. H. & Pintado, M. R. (1985) The wear of dental amalgam in an artificial mouth: a clinical correlation. *Dent. Mater.* **1** 238–242.

DeLong, R., Douglas, W., H., Sakaguchi, R. L. & Pintado, M. R. (1986a) The wear of dental porcelain in an artificial mouth. *Dent Mater.* **2** 214–219.

DeLong, R., Sakaguchi, R. L., Douglas, W. H. & Pintado, M. R. (1986b) Remote-site measurement of occlusal wear. *J. Dent. Res.* **65** 749.

DeLong, R., Douglas, W. H., Pintado, M. R. & Sakaguchi, E. L. (1988) Accuracy of 3-D digitization of posterior composite wear. *J. Dent. Res.* **67** 362.

Douglas, W. H., Sakaguchi, R. L. & DeLong, R. (1985) Frictional effects between natural teeth in an artificial mouth. *Dent. Mater.* **1** 115–119.

Morin, D. L., Douglas, W. H., Cross, M. & DeLong, R. (1988a) Biophysical stress analysis of restored teeth: experimental strain measurement. *Dent. Mater.* **4** 41–48.

Morin, D. L., Cross, M., Voller, V. R., Douglas, W. H. & DeLong, R. (1988b) Biophysical stress analysis of restored teeth: modelling and analysis. *Dent. Mater.* **4** 77–84.

Nelder, J. A. & Mead, R. (1965) A simplex method for function minimization. *Computer Journal* **7** 308–313.

Pintado, M. R., Conry, J. P. & Douglas, W. H. (1988) Measurement of sealant volume *in vivo* using image-processing technology. *Quintessence International* **19** 613–617.

Sakaguchi, R. L. (1988) A biophysical analysis of the occlusal wear of dental materials. PhD thesis.

Sakaguchi, R. L., Douglas, W. H., DeLong, R. & Pintado, M. R. (1986) The wear of posterior composite in an artificial mouth: a clinical correlation. *Dent. Mater.* **2** 235–240.

9

Surfaces in solid modelling

M. J. Pratt
Professor of Computer Aided Engineering, Cranfield Institute of Technology

The types of surface representation commonly used in solid modelling are surveyed, and some of the geometric computations required in a practical system of this type are described. One particular modelling problem, the automatic generation of blends, is then examined in more detail, and two new approaches to blending are outlined. The object of these methods is to provide modelling techniques which are totally intuitive for the user, easy to implement, largely free of requirements for complex and intensive geometric computations, and consequently rapid in response. The chapter concludes with a plea for more research in the development of improved user interfaces in solid modelling.

1 INTRODUCTION

Throughout most of the history of CAD systems, now spanning nearly thirty years, there have been two distinct strands of development. The first started with the implementation of 2-D draughting systems, which gave rise in the 1970s to 3-D wireframe systems and more recently to solid modelling systems. The second, whose origins may arguably be traced back to methods used by shipbuilders in antiquity, has led to today's systems for free-form curve and surface definition. There has indeed been some interaction between the two schools of development over the past decade or so, but their full integration has proved difficult to achieve. A primary reason for this is that the types of geometry generally employed in the two classes of system have significant differences. In particular, certain types of geometric calculation involving free-form curves and surfaces present major computational challenges.

Draughting and wireframe systems usually function in terms of 'traditional' curve geometry such as straight lines, circles, and other conic curves. Where surfaces are implemented, particularly in solid modellers, they are usually planes and the other 'natural' quadrics (spheres, right circular cylinders, and cones). The torus, a non-quadric surface, is also often provided for dealing with configurations such as pipe elbows and fillets between planes and cylinders.

The free-form curve and surface systems in use for the design and manufacture of aircraft, cars, ship hulls, shoes, plastics bottles, and so forth are on the other hand based upon different concepts. Both the underlying constructional ideas and the theory of parametrically defined curves and surfaces have a long history, but it was not until the advent of powerful computers in the fifties and sixties that the two could be combined in the powerful methods used today. The invention of polynomial splines in the forties also had a strong influence on this development.

This chapter will survey some aspects of the current state of integration of free-form methods into solid modelling. Firstly, the essential differences between the two types of geometry used will be outlined; interrelationships between them frequently exist, however (Pratt 1990). Secondly, a brief discussion will be given of some of the difficult mathematical problems arising from the implementation of free-form surfaces into solid modellers. Thirdly, some often neglected aspects of user interface design for modelling systems will be examined. Fourthly and lastly, some of the previous material will be illustrated in terms of some new ways of tackling the blending and filleting problem in solid modelling.

2 IMPLICIT AND PARAMETRIC GEOMETRY

Equations of the form $f(x,y)=0$ and $f(x,y,z)=0$ are known as implicit; they represent respectively a curve in two dimensions and a surface in three dimensions. The corresponding parametric representations of curves and surfaces have the form $r=r(t)$ and $r=r(u,v)$ respectively, where r is the position vector of a point. These matters are well-known and are explained, for example, in Faux & Pratt (1979). Note that the parametric curve, unlike the implicit one, may be a 3-D space curve. To represent such a curve in implicit terms requires the definition of two implicit surfaces, whose intersection gives the desired curve.

Implicit equations define curves and surfaces which are unbounded, unless some restriction is placed on the range of the relevant variables. Thus the equation $ax+by+cz=0$ represents a plane of infinite extent, and $x^2+y^2=0$ represents a circle which, although having a finite circumference, is not a bounded curve. By contrast, parametric curve segments used in CAD are defined in terms of some bounded range of their parameter, often $0\leq t\leq 1$, while surface patches are similarly defined in terms of bounded ranges of both parameters. The chief reason for this is that such geometric entities are often effectively specified by the values of certain quantities on their boundaries (i.e. at the endpoints of a curve segment or the corners of a surface patch). This approach can be used to give good control over the interior of the segment or patch, but it gives very little control over what happens outside the chosen parameter range. In fact if the parameters are allowed to vary outside this range, all manner of wild behaviour results, including cusps and self-intersections in the case of curves, and cusps, sharp ridges, and self-intersections in the case of surfaces. Parametrically defined geometry is therefore not unbounded, as is frequently imagined, but for practical purposes has bounds imposed upon it. The reason for not treating implicitly defined geometry similarly is that the surfaces commonly used in CAD are reasonably well-behaved as the variables tend toward infinity.

The following subsections give brief descriptions of the way in which the two types of representation are used in typical modelling systems. Readers unfamiliar with solid modelling may refer to Rooney & Steadman (1987), Woodwark (1986), Mäntylä (1988), or Chiyokura (1988) for further details.

2.1 Solid modelling — constructive solid geometry

Constructive Solid Geometry (CSG) is an approach to the computer modelling of solid forms by building them up from simpler volumetric elements called 'primitives'. The primitives provided in typical CSG systems include rectangular blocks, cylinders, cones, spheres, and toruses. These are combined by using what are known as 'Boolean operations', which are the set operations of union, difference, and intersection applied to the volumetric primitives regarded as sets of points in space. Thus an L-shaped bracket may be modelled as the union of two blocks, a hole then modelled by differencing or subtracting a cylinder, and so on.

Conventionally, the primitive volumes in a CSG system are represented in terms of 'half-spaces'. Two half-spaces are defined by any surface which divides the whole of 3-D space into two disjoint regions. Thus a plane gives rise to two half-spaces, one on either side of it; an infinite cylinder similarly gives rise to two half-spaces, one inside and one outside the cylinder. A half-space is not necessarily infinite in volume, as in these two cases; a spherical surface defines an inner and an outer half-space which are respectively bounded and unbounded in volume.

The surfaces in CSG systems are defined implicitly, by equations of the type $f(x, y, z)=0$. This enables the half-spaces defined by each surface to be distinguished from each other as the sets of points for which $f<0$ and those for which $f>0$. For example, the inner half-space of a vertical cylinder of radius 3 is defined by $x^2+y^2<9$. To turn this into a bounded volumetric primitive, a cylinder with finite length, two planar half-spaces are needed, for example $z<6$ and $z>0$. The set intersection of the three half-spaces then results in a set of points for which $x^2+y^2<9$ and $z<6$ and $z>0$; this set of points occupies a cylinder of radius 3 and finite length 6. A rectangular block is similarly defined, as the set intersection of six planar half-spaces.

The basic data structure of a CSG modeller gives very compact representations of even quite complex solid shapes in terms of a sequence of primitives and set operators, which may alternatively be expressed as a graph structure known as a 'CSG tree'. Since the main consideration in this chapter is surface geometry no further details will be given here; more information may be found in Rooney & Steadman (1987).

2.2 Solid modelling — boundary representation

A 'boundary representation' model of a solid form may be distinguished from a CSG model by the type of data structure employed, which contains much more detail and often, for reasons of computational efficiency, a high degree of redundancy. In its simplest form the data structure takes the form of a network whose nodes represent objects, faces, edges, and vertices. An object node as pointers (links) to a set of face nodes, representing all the faces forming the boundary of the object which divides its interior from its exterior. Each face node has pointers to a set of edge nodes, each of which in turn has pointers to two vertex nodes. Pointers in the reverse directions are also usually implemented. The structure so far described is purely 'topological', since it merely specifies how a set of faces, edges, and vertices are connected together. The object represented may be thought of as a deformable shape, which does not become rigid until geometry is associated with the topological elements in the network. Thus in the final model each face node will have fruther pointer to a surface node, each edge node to a curve node, and each vertex node to a point node. Note that the curve and point nodes are redundant, since the curve of an edge may always be computed as the intersection of the surfaces of the two faces meeting at that edge; similarly, the point of a vertex is the intersection of three surfaces. However, it is convenient for many purposes to have detailed geometric information concerning the edges and vertices present in the model, and so this information is in most boundary representation modellers computed during creation of a model and stored explicitly for further use. The description given here is simplified in several respects from the data structures used in practical systems, but the essential principles should be clear. Again, further details are given in Rooney & Steadman (1987).

Surfaces in boundary representation solid modellers may be modelled in several ways:

(1) Some systems use implicit representations as described above for surface representations in CSG modellers. This has the advantage that a label can be attached to indicate whether the inside of an object lies in the positive or the negative half-space associated with any face.
(2) Alternatively, some systems exclusively use parametric surfaces of the type discussed in the next subsection; the problems of this approach will be examined there.
(3) Thirdly, some systems define the geometry of curved surfaces approximately in terms of assemblies of planar or near-planar facets. The advantages and disadvantages of this approach will also be discussed later, in subsection 2.4.

There are also other possibilities; for example, Goldman (1983) points out the virtues of what he calls 'geometric' representations of surfaces. He suggests, for example, that a cylindrical surface should be defined simply in terms of its axis and its radius rather than by an equation of the type given above in section 2.1. Furthermore, many practical solid modelling systems do not conform exactly in their methods of surface representation to the three alternatives listed above; some use hybrid methods which combine two of these approaches.

2.3 Surface modelling

Solid modelling systems provide various means of checking that models are valid in the sense of topological completeness of their boundaries. Pure surface modellers, however, make no pretence at this. They are intended for the modelling of single surfaces, or of assemblies of surfaces, but make no special provision at all for the modelling of the exteriors of complete solid objects. If this is done with such a system, validity of the result is entirely the user's responsibility.

Surface modellers use parametrically defined geometry, usually based on the use of polynomial or rational functions of the parameters. If a surface is defined by $r=r(u,v)$ then the tangent vectors of the isoparametric curves at any point with parameter values u,v are $r_u(u,v)$ and $r_v(u,v)$. These two vectors define the tangent plane of the surface at the given point provided that their cross-product is nonzero. If it is zero then either the isoparametric lines are parallel at the point in question or one of the tangent vectors has zero magnitude there, which probably implies the existence of a sharp cusp or ridge in the surface (Faux & Pratt 1979). These unusual and undesirable cases will here be neglected, and it will therefore be assumed that the cross-product defines a non-null vector normal to the surface. Unfortunately this does not imply that a parametric surface is orientable as are the implicit surfaces discussed above. In that case the orientation is a specific property of the surface, whereas in the parametric case it is easy to find reparametrizations of the surface which reverse the direction of the surface normal and correspondingly the sense of 'inside' and 'outside'. For example, if the parameter ranges are $0 \leq u \leq 1$ and $0 \leq v \leq 1$ then the simple parameter change $u' = (1-u)$ leaves the geometry of the surface unchanged but reverses the direction of the surface normal everywhere. Unlike an implicitly defined surface, then, a parametric surface does not have an inside/outside orientation, and it is not directly possible to use such surfaces in CSG modelling systems which use algorithms based upon half-spaces. Recall also that it is only bounded regions of parametric surfaces which are used in practice, and these clearly cannot define half-spaces. Any attempt to overcome this limitation by permitting infinite parameter ranges runs into problems due to the unpredictable behaviour of parametric geometry outside the range on which it was originally defined; self-intersecting surfaces, for example, present obvious difficulties in a half-space context. For these reasons it is proving difficult to implement general free-form or sculptured surfaces in a CSG context (Dunnington *et al.* 1989). The problems of implementing them in a boundary representation system are less; a small number of commercially available solid modellers already provide a combination of implicit and parametric surface types, though the most detailed account of such a system in the literature of the subject concerns a research modeller (Várady 1985).

2.4 Faceting systems

Some of the practical problems which arise when sculptured surfaces are implemented in a solid modelling context are discussed in the next section. Several commercial system developers have attempted to avoid these by approximating such surfaces as assemblies of planar (or near-planar) facets. To take just one example, the calculation of the intersection curve between two general parametrically defined sculptured surfaces is a major computational exercise. The faceting philosophy is that this single complex calculation can be reduced to a collection of plane/plane intersection calculations, which are individually very much simpler. There are penalties, of course; for one thing it is possible that the incorrect object topology may result from such a calculation, and for another the use of facets is not appropriate when precise dimensions and tolerances are of importance, as they are in many engineering applications. To alleviate these problems the precision of the approximation may be enhanced by increasing the number of facets, but here there is a trade-off, since if the increase is beyond a certain level the intersection computation will require so many plane/plane calculations that it would be quicker to solve the single complex intersection problem in terms of the exact geometry. In fact most of the developers of faceting solid modellers are now working on the provision of algorithms based on the use of exact surface representations. For this reason, despite its current prevalence, the faceting approach will be ignored in the remainder of this paper.

3 SURFACE INTERROGATIONS

In this section some of the computational requirements of sculptured surfaces in solid modelling will be briefly examined. Most of the topics mentioned are covered in more detail in Pratt (1990), where a more complete list of references will also be found.

3.1 Surface intersections

In a solid model an edge is topologically a set of points shared by two adjacent faces; geometrically, it is a bounded segment of the curve of intersection of the surfaces on which those faces lie. It is therefore important to be able to compute such intersection curves. There are well-established methods for doing this for the simpler implicit surfaces used in solid modelling, particularly the quadrics (Levin 1976), but efficient and robust methods for the intersection problem for general parametric sculptured surfaces are still being sought. This important area of research is surveyed in Pratt & Geisgow (1986). The reader should note that while edge curves are solutions of the two-surface intersection problem, vertex points lie at the intersections of three surfaces.

3.2 Trimmed surfaces

If a face of a solid model lies on a parametric sculptured surface its edges will not generally lie on the boundaries of the surface. This is to say that the face is a bounded region of the overall surface, and the question arises as to the nature of the bounding mechanism. So far as solid modelling is concerned attention will be confined to the boundary representation approach, since there are few if any effective sculptured surface implementations in CSG systems. In this case the bounding will generally be topological; the face will have a pointer to the underlying surface and to a set of edges. These in turn will have pointers to the curves on which they lie — often intersection curves computed as indicated above. The face may be quite complicated and have multiple boundary loops (if it contains internal holes, for example). Discussions of the problems of trimming surfaces may be found in Farouki (1987), Casale (1987), and Hook & Tiller (1989).

Some pure surface systems also provide facilities for defining trimmed surfaces, but added difficulties arise if the systems do not model topology. In such a case the boundary curves of the trimmed region can be computed and (at least in the simplest case of a four-sided region) some Coons-based technique used to define a new surface patch which interpolates them (Faux & Pratt 1979). Clearly this will be only an approximation to the original trimmed region so far as the interior of the region is concerned. More complex regions must be subdivided into four-sided subregions for this technique to be used, and more sophisticated approximation methods than the simple Coons approach may take into account interior as well as boundary data to obtain better agreement with the original surface. Some of the approximation methods cited in section 3.4 below may be used for this last purpose.

3.3 Silhouette lines and other graphical interrogations

Graphical interrogations are used for generating pictorial representations of an object, and are viewpoint-dependent. For example, 'silhouette' or 'profile lines' are visual edges of the object depicted, the locus of points at which lines through the viewpoint are tangent to its curved surfaces. Whereas the calculation of silhouette lines is chiefly a requirement in the context of line drawings, shaded surface renderings require consideration of the angle at which any line from a source of illumination strikes the surface, and also of the angle made with the surface by the line of sight. Here the computation of the surface normal direction is of primary importance. Such graphical interrogations are discussed in detail in standard texts on computer graphics, e.g. Foley & van Dam (1982).

3.4 Data exchange problems

It is found increasingly necessary in an industrial context to be able to transfer CAD data between different systems. This may be a requirement in a large company using several different systems for diverse purposes, or it may be that a primary company wishes to transmit data to a subcontractor who uses a different system. Clearly it is inefficient to make such transfers by sending drawings and having the geometry input manually into the receiving system; what is required is a transfer on some medium such as magnetic tape and the automatic reconstruction of the model in the receiving system. One major problem concerned with sculptured surfaces is that systems use widely different representations for such surfaces. Polynomial systems range in degree from two to twenty-six; other systems use rational representations of various degrees, and yet others use procedural representations in which surfaces are defined in terms of a set of boundary curves and a blending procedure. It is clearly not possible to transfer data representing a surface of degree 20×20 exactly into a system which can handle only bicubic surfaces (though the reverse transfer is of course possible). Similarly, exact transfer from a rational into a polynomial system is not in general possible, though once again the reverse transfer may be made, subject to a proviso on the degree of the functions concerned, since the polynomials are a subset of the rationals.

In cases where exact transfer is not possible it is necessary to use approximation techniques. Usually some tolerance will be imposed upon the accuracy achieved. The methods of classical approximation theory prove not to be suitable without modification, and two approaches to suitable methods for CAD purposes can be found in Lachance (1988) and Goult (1989).

3.5 Implicitization and parameterization

Little will be said about this topic, which is concerned with the determination of implicit representations of parametric surfaces and vice versa. It has only recently been realized that the former type of conversion is possible (Sederberg *et al*. 1984), and some of the methods used are based upon hitherto neglected algebraic techniques developed in the Victorian era. Such conversions are potentially of importance in the computation of intersection curves (see section 3.1 above), since certain simplifications occur when one of the surfaces is expressed in implicit and the other in parametric form (Pratt & Geisow 1986).

3.6 Automatic machining of surfaces

Just one manufacturing application will be mentioned, the automatic generation of control information to drive a numerically controlled machine tool used for machining a surface. This is simplest when the surface is parametrically defined and the cutting tool has a ball or spherical end. In this case the centre or reference point of the sphere moves on a surface which is a constant parallel offset from the original. Points on this offset surface are easy to calculate; at any point on the original surface the cross-product of the tangent vectors along the lines of constant u and constant v, provided that it is non-null, defines a vector normal to the surface. It is only necessary to move in this direction by the offset distance d to generate a point on the offset surface. Matters become much more complicated if the geometry of the cutting tool is more general, but the calculation of the positions of tool reference points as the cutter moves on the surface follows the same principles. Still more complex types of calculation are often required in which the cutting tool moves so that it is simultaneously in contact with two surfaces. Some further details of calculations of this kind are given in Faux & Pratt (1979).

3.7 Geometric queries

It should be noted that tangent and normal vectors to the surface are needed in the calculation of shaded surface renderings and the generation of machining data. These are first derivative properties, which are also needed for the calculation of the lengths of curves and the areas of bounded regions on the surface. Second derivative properties are related to curvature , and these are also important in some applications. For example, if it is desired to machine a concave surface with a ball-end cutter, the radius of the cutter must be smaller than the minimum radius of curvature of the surface, otherwise unwanted gouging will occur. The intention of this brief section is merely to point out that computation of the elementary geometric properties of surfaces is a necessary prerequisite to most of the surface interrogations listed above. Further details on this topic are given by Faux & Pratt (1979), Nutbourne & Martin (1988), or Pratt (1990).

4. USER INTERFACE PROBLEMS

In a paper by Várady & Pratt (1984), various approaches to the creation of solids with sculptured faces are classified from the user's point of view. The methods were categorized as follows:

(1) Surface/solid operations — in this group of techniques a solid model is assumed to exist initially. An isolated sculptured surface is then created and used to modify the original solid. In the simplest case the surface is defined so that it cuts the solid into two portions; one of these is then rejected, leaving a smaller solid having a sculptured face. The original paper describes a number of variations on this basic theme.

(2) Creation of free-form solid primitives — here again an isolated sculptured surface is created, which may be used in various ways to define a solid primitive for use in subsequent Boolean operations. For example, a simple four-sided surface patch may have its edges projected in a specified direction onto a plane to

create a sculptured solid. Alternatively, if the surface was created by translating a closed profile along a curve or by interpolating a set of closed profiles a duct-like surface results which can be made into a solid simply by closing off the ends. Another method of creating a solid from a bounded surface is by simply adding thickness to the surface.

(3) Automatic blending — this technique is currently the subject of much research, because it has been identified as a key requirement for the modelling of final detail in many engineering applications. The user would like to be able to indicate any sharp edge on the model and simply request the system to round it off, that is construct a blend or fillet between the two surfaces meeting at the edge in question. Typically he would also wish to specify a blend radius, or perhaps two radii, one at each end of the edge to be blended if a non-uniform rounding is desired. Some of the problems of this type of operation will be examined in more detail in section 5 below.

(4) 'Rubber face' operations (referred to in the original paper as 'rubber object' methods) — this class of operations requires the preexistence of a solid model. The geometry of one or more faces of the model is then 'unfixed' in such a way that it can be manipulated by the user to achieve some desired sculptured configuration. This type of operation will also be examined more closely in section 5. In the original paper by Várady & Pratt a fifth class of methods was also defined, 'Creation of new free-form faces on an existing object', but these techniques are very closely related; they require the initial creation of a bounded region of an existing object face followed by a 'rubber face' modification of that region.

The four classes of methods fall very clearly into two groups. (1) and (2) both require the creation of a surface in isolation, while (3) and (4) do not. The groups may be referred to as 'surface-based' and 'solid-based' methods respectively. Some of their characteristics are examined below.

4.1 Surface-based methods

Surface-solid operations require the user to create a surface and to position it appropriately with respect to the existing solid. The sculptured object created by these operations will in general have sharp edges bounding the sculptured region of its exterior; these edges will have been computed as bounded segments of intersection curves between the original sculptured surface and the relevant faces of the original solid. The action of this technique therefore requires significant computation by the system, and the result created is untypical of many engineering objects, in which sculptured faces blend smoothly into other faces. Any attempt to achieve such blended results by using the surface–solid technique will require the creation of an original surface which is tangential to the solid on some of its faces. The computation of the intersection curves then becomes extremely hazardous, since rounding errors could perturb the geometry of the surfaces concerned sufficiently that no intersection or line of tangency could be found to exist.

The solid primitive technique also has the disadvantage that it creates solids with sharp edges to the sculptured portion of their exteriors. Further, although the initial creation of such solids is computationally inexpensive, their use in further modelling operations often requires the use of Boolean operations, which are very expensive.

It may be concluded that neither of the surface-based classes of method has very general application for the realistic modelling of engineering parts, and that both imply the use of computer-intensive intersection or Boolean calculations.

4.2 Solid-based methods

There is an overlap between these two classes of method, in that one approach to automatic blending is based on the use of rubber face techniques. A key characteristic of both classes is that a preexisting solid model is used as a framework for the construction of the sculptured geometry. It is not necessary for the user to create, position, and orient a surface which is originally not connected to the solid model; in these methods the defining data, the position and orientation of the new surface, are all determined directly and automatically from the geometry of the solid model. A further virtue is that these methods are perfectly capable of defining the smooth blends between faces which the surface-based methods cannot handle. This is because the edges of blend surfaces are known before the surfaces are constructed, and because the system is aware of requirements for tangency between surfaces and can take them into account in the determination of the new geometry. Furthermore, many applications of these methods require no surface intersection or other intensive calculations. Although such techniques are difficult to implement in a CSG context, in a boundary representation system the bounded surface regions which have been determined can be interpolated into the model data structure by using what are called 'local operations', which are very efficient since they involve only minor local topological changes (Braid 1979, Mäntylä 1988).

Although the solid-based approach to the definition of free-form geometry has the advantages stated, methods of this type have been little implemented with the exception of algorithms of the automatic blending class, and even these are still mainly in the research stage. The wider use of such methods will have the virtue not only of making solid modellers more efficient (both from the point of view of wider geometric coverage and of less computationally intensive operations) but also of making them easier for the user to work with.

5 THE BLENDING AND FILLETING PROBLEM

The aim of this penultimate section is to illustrate some of the foregoing points in terms of current research with which the author is associated. First, however, the nature of the blending problem in solid modelling will be examined in more detail, and some earlier approaches to its solution described.

In solid modelling terms a 'blend' or 'fillet face' is a bounded region of a surface providing a smooth transition between two other surfaces which would otherwise intersect in a sharp edge. The word 'smooth' here implies continuity of direction of the surface normal vector along any path across the boundary between the blending face and either of the other faces.

Both approximate and exact blending methods have been proposed. Approximate techniques can be further subdivided into those using faceting and those using parametric surface strips. Faceted approximations are clearly unsatisfactory, not only for the reasons outlined in section 2.4 above but because they violate the notion of smoothness inherent in the definition of a blend.

The parametric strip type of approximation is a great improvement. In a typical implementation the boundaries of the blends are first calculated; one method of doing this is to compute the intersection curve of an offset of the first surface with the second surface, and vice versa. Tangent directions of the first and second surfaces transverse to the boundary curves may now be calculated along their lengths, and the positional and tangent information together used to construct a strip of parametric surface patches which has the required properties. In fact the curves and the tangent values can usually be computed only at discrete points rather than continuously, so that the resulting blend is only approximately continuous in position and tangent across its edges (Várady *et al.* 1989b).

Most approaches to the definition of exact blends use a method of algebraic combination of two implicitly defined surfaces intersecting in a sharp edge. Several authors have recently suggested such techniques, but only one commercial implementation is known at present (Rockwood & Owen 1985). An interesting and novel alternative has been implemented by Chiyokura (1988), though his initial model is of the wireframe rather than solid type. When an edge is to be blended, certain topological and geometrical modifications are made to the wireframe, consistent with the presence of a blending face to replace the edge. The last step is the automatic generation of faces, including blending faces, to clothe the wireframe model completely and transform it into a solid model of the boundary representation type. Yet another approach, this time based on a wireframe approximation to the object modelled, treats the wireframe as the control net of a smooth *B*-spline surface, and uses recursive subdivision to generate that surface (Veenman 1982, Nasri 1984).

Good reviews of the various blending algorithms are given by Woodwark (1987) and Várady *et al.* (1989a).

The most natural mode of interaction for the user in creating a blend is to point to an edge, perhaps specify two or three numerical parameters to define the precise nature of the blend, and to allow the system to do the rest of the work. Most of the methods mentioned permit this mode of operation with the exception of those using faceting; these are highly prone to failure because of the likelihood of facets belonging to the original surfaces and facets belonging to the blend surface actually being parallel in the neighbourhood of the blend boundaries. In such a case these boundaries, which the system is trying to compute as a sequence of plane/plane intersections, will fail to exist in their entirety, and the modeller is likely to crash.

The non-faceting methods described are reasonably successful for the blending of single edges, or in some cases for two or three edges meeting at a vertex, which will normally also be appropriately blended to avoid sharp corners. However, more complex blending situations may give rise not only to difficult geometric problems (for example the treatment of overlapping blend surfaces) but also to significant topological difficulties (Várady *et al.* 1989b). Further work is needed to find ways of dealing with these complexities.

The two new blending methods described in what follows suffer from the same limitations as have just been described, but each is felt to offer significant advantages in certain relatively simple situations. Again, further research may widen their applicability.

5.1 Cyclide blending

Cyclides are surfaces which may be thought of as generalizations of the torus in which the minor radius varies around the central hole. They were first discovered by the French mathematician Dupin (at the age of 17) in 1801, but found few applications until their potential use in CAD was studied in Cambridge during the present decade (Nutbourne & Martin 1988). The Cambridge work was aimed at the use of piecewise cyclides in free-form surface modelling with controlled curvature characteristics; subsequent research by the present author has shown that these surfaces may be used to provide a limited blending capability in a solid modelling context (Pratt 1989a, b). A key property of the cyclide is that it may be parameterized in such a way that the two sets of isoparametric lines are both families of circles. Furthermore, members of the two families intersect orthogonally. It is shown in the papers referred to that any one of these circular isoparametric lines is a line of tangency with a certain right circular cone. Then if we can find two cyclides which are both tangent to a meridional circle on a particular cone they are also tangent to each other, and they blend smoothly one into the other.

At this point it is relevant to note that the surfaces conventionally used in solid modelling are the plane, circular cylinder, cone, sphere, and torus, which all happen to be special cases of the cyclide. It is therefore possible to blend smoothly from any one of them into a cyclide surface, and then to blend from that cyclide surface into a further cyclide, and so on. Some simple cases are illustrated in Fig. 1. Certain slightly more constrained situations require the use of a 'double-cyclide blend,' made up of portions of two cyclides (see Fig. 2).

From the system developer's point of view the implementation of cyclide surfaces is not difficult. Like the torus, which is almost universally provided in modern systems, they have a fourth-degree implicit representation. The computation of intersection curves with a general cyclide, to take just one example of a geometric interrogation, is very little more difficult than for the torus. The provision of the more general surface, however, leads to a very useful increase in the range of blends which can be modelled exactly. There is a severe limitation of the method as described, in that the blend boundary curves must always be circles. Boehm (1989) has described an extension of the technique which has promise for the construction of more complex blends, and the present author is studying generalizations of cyclides in an attempt to identify a wider class of useful blending surfaces.

As regards the earlier classification of surface definition techniques in solid modelling, this method is an automatic blending method. The preexisting, unblended solid model is used as a framework. Once the user selects the edge to be blended, the system can identify the surfaces meeting at that edge and can determine whether they are both cyclides or special cases of cyclides. If so, then a family of cyclide or double-cyclide blending surfaces will exist. The user can choose one member of this family by specifying a single parameter (for example a minimum or a mean radius for the blend), and the system can then automatically determine the appropriate particular blend surface together with the transformation needed to position and orient it to suit the required geometric context. The method is therefore very simple for the user, and gives a result in terms of exact geometry for a range of restricted but commonely occurring situations.

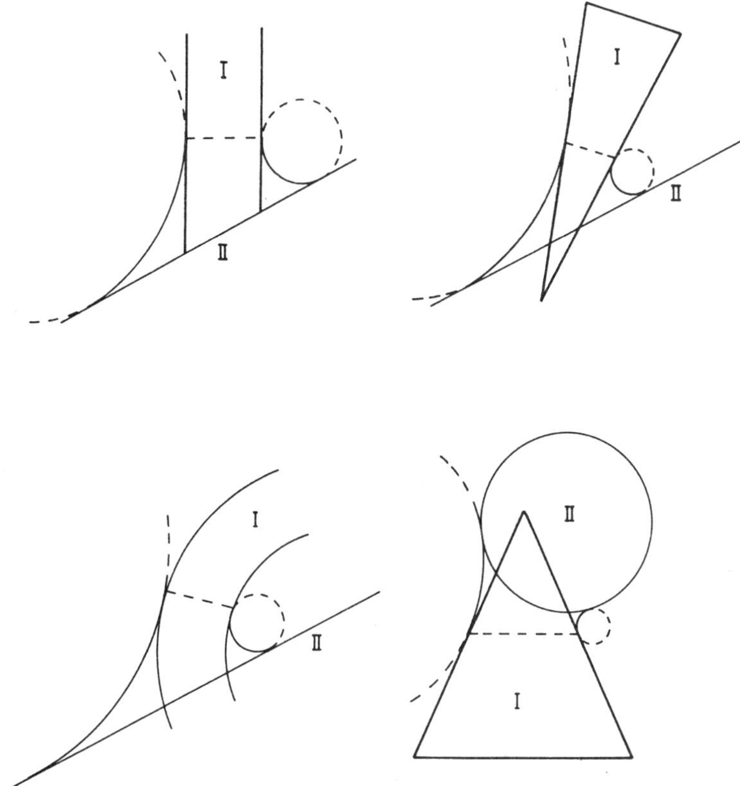

Fig. 1 — Cross-sections of simple cyclide blends in their planes of symmetry, in which the cyclide cross-sections are two circles. The cases shown (I/II) are cylinder/plane, cone/plane, torus/plane, and cone/sphere.

5.2 Rubber face blending

The potential advantages of methods of this type were pointed out by Várady & Pratt in 1984. Subsequently Pratt (1985) described a rubber face method of automatic blending, but not until recently has this been tried in practice. The approach is designed to use well-known geometric techniques in a way convenient for the designer.

Consider first a rectangular block, modelled in standard boundary representation form in terms of a set of planar faces with their associated edges and vertices. Normally the planar surface of each face will be defined by an implicit equation or some closely related form. Now suppose it is desired to modify the top face to give it some sculptured configuration. To this end, the top face may be redefined in the model as a Bézier patch; in the bicubic case this simply requires the creation of sixteen control points all lying in the original plane, four at the corner vertices, two interior to each edge, and four in the interior of the face. At this stage the geometry of the face is unaltered; the only modification to the model is a change from an

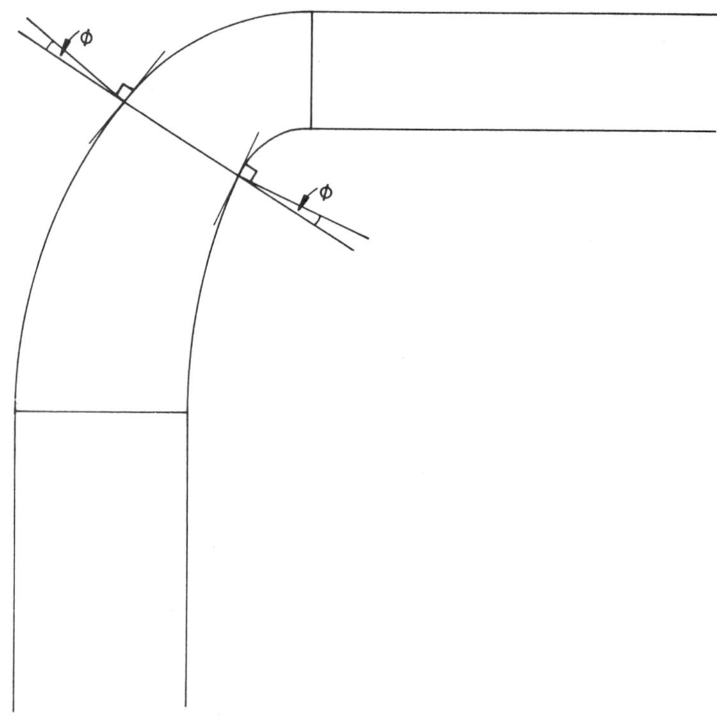

Fig. 2 — Double-cyclide blend between two pipes of unequal diameters. The two cyclides are mutually tangent on a circle, around which the surface normal makes a constant angle with the indicated plane.

implicit to a parametric representation of one surface. The four interior control points may now be manipulated by moving them out of the plane. This will not affect the linear boundary curves of the force, but its interior will depart from planarity and assume some sculptured shape. Note that this has entailed no topological change in the model whatsoever, and no significant geometric computation.

Automatic blending requires similar operations on two or more faces. Suppose now that it is wished to blend the top front edge of the block. The first step is to redefine both the top and the front face of the block as Bézier faces, of the same degree. If these are bicubic as before, then there will be a shared set of four control points lying in a straight line along the edge to be blended. If these four points are moved simultaneously so that each lies midway between the corresponding neighbouring control points interior to the top and the front face, then the properties of Bézier surfaces ensure that a smooth blend will occur between the modified faces. Again, no topological changes have occurred and no intersection or other intensive geometric calculations have been required. The resulting geometric changes, apart from those in the faces in question, are (1) changes in four edge curves — those of the left and right edges of the top and front faces, and (ii) changes in two vertices — those of the edge which has been blended. No additional computation is needed to elucidate these changes, however, since the new edges simply lie on boundaries of the modified faces (and are consequently given by the control points controlling those boundaries), while the new vertices coincide with corner control points shared by the modified faces. Note that the shape of the side faces has changed, but that this shape is determined by the modified edge curves; there is no change in the surface geometry of these faces.

It is possible to extend this approach to the blending of two or of three edges meeting at a vertex, as discussed in Pratt (1985), which also shows how a cylinder may be blended smoothly into a planar face and suggests how more complex blending cases may be dealt with. The two- and three-edge cases must be based on a careful mathematical analysis of the continuity conditions between three Bézier patches meeting at a common vertex rather than the four which is standard in a conventional piecewise parametric surface made up of four-sided patches. Bézier himself (1986) has recently provided such an analysis. The blending of cylinders requires the use of rational rather than simple polynomial representations. An experimental implementation of some of the cases described has now been made, and Figs 3 and 4 show some

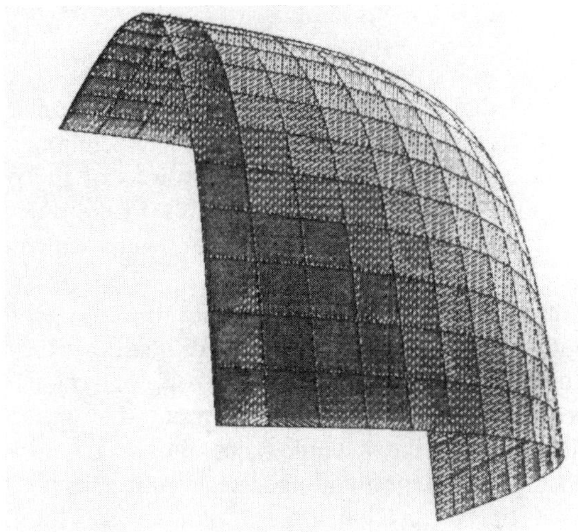

Fig. 3 — Rubber face blend of three edges of a cube and their associated vertex.

of the blends which have been achieved. The details will be more fully reported elsewhere.

In terms of the classification given earlier, the blending technique described is both a rubber face method and an automatic blending method. The implementor needs to provide automatic transformation of implicit to parametric surface geometry as required, and to make available a means for the manipulation not only of single control points but also of lines and groups of control points. Both are relatively simple to achieve. Control point manipulation may or may not be explicitly available to the user; in the simple edge blending described above, for example, the entire

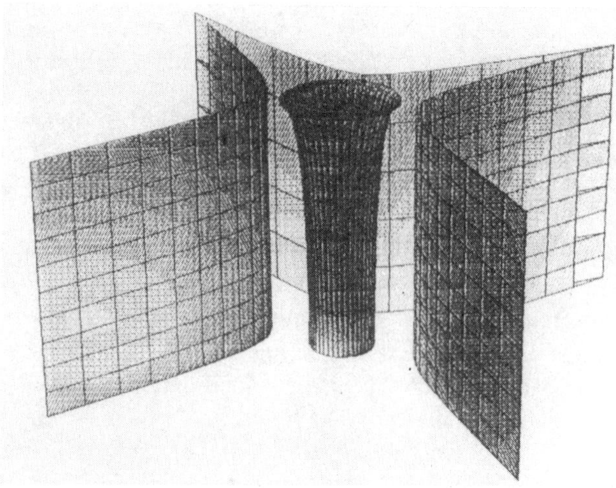

Fig. 4 — Ribber face blend of a cylidrical hole onto the top face of a block. Rubber face changes have also been made to three side faces of the block; the remaining planar faces are not drawn.

process can be made automatic once he has indicated the edge to be blended. On the other hand, some knowledge of the principles of the Bézier representation of curves and surfaces will allow all the power of the rubber face technique to be available to him, including a wide range of possible generalizations of the particular method described in the previous paragraphs. Note that at no time does the user need to position and orient surfaces in space; this is all done automatically with respect to the previously defined object geometry.

Although the rubber face approach to the generation of free-form surfaces in a solid model has great potential, it is not without its difficulties. Rubber face blending, in common with other blending methods, runs into topological problems in a range of commonly occurring situations where there are multiple edges. Furthermore, the technique as described above affects whole faces, though the greatest effect occurs near the edge or edges specified for blending. The blending may, however, be strictly localized in effect by initial subdivision of the original faces through interpolation of new edges parallel to the one to be blended. Then it is only the subfaces adjacent to that edge which are modified, while the others remain unchanged. Some subdivision technique of this kind will in any case be needed when a face is being dealt with which is not four-sided, since such a face cannot be expressed in the form of a single biparametric patch.

6 CONCLUSIONS

In the first part of this chapter the types of surface representation used in solid modelling have been described. Some of the geometric interrogations necessary for the implementation of practical modellers have then been briefly surveyed, and some aspects of user interface design discussed. Problems specific to the generation of blending surfaces in a solid modeller have been outlined, and finally a description has been given of two new approaches to blending, aimed at avoiding some of these problems.

There is much current research into improved geometric algorithms for surface intersection and similar calculations. Much less emphasis is given (at least in the published literature) to improving the user interfaces of geometric modelling systems to allow their efficient use by mathematically unsophisticated designers. This paper is a plea for researchers and developers to give more attention to this aspect of the field. If at the same time ways are sought of avoiding the necessity for complex geometric interrogations, matters will be still better for the user, since the system response time will improve. At present slow response is one of the chief bugbears of solid modelling systems. The blending methods described provide indications of two possible ways in which these aims can be achieved; both of them make use of previously known surface geometry, are comparatively simple to implement, avoid intensive geometric calculations, and are easy for the user,

There are probably many other such methods waiting to be discovered, some doubtless better than those presented here. We now have a great deal of information at our disposal concerning the mathematics of surfaces, and one of our major problems seems to be to make the full power of this knowledge available to the nonmathematical user in such a way that he can rapidly gain an intuitive understanding of it. Essentially, the requirement is for the user to operate in what is to him a totally natural way, while the system decides what is the appropriate type of underlying mathematics to use. The situation at present is that the user often has to make this decision and then use different techniques according to the chosen type of geometry. This is clearly far from an ideal situation, and it is hoped that the present chapter may stimulate further user-oriented research.

REFERENCES

Bézier, P. (1986) *The mathematical basis of the UNISURF CAD System*, Butterworths.

Boehm, W. (1989) Extensions of cyclide blending, paper presented at conference on Mathematics of Surfaces, Oberwolfach, F. R. G., April 1989, *Computer Aided Geometric Design* (in press).

Braid, I. C. (1979) *Notes on a geometric modeller*, CAD Group Document 101, Cambridge University (revised 1980), now obtainable from DACAM, Cranfield Institute of Technology, England.

Casale, M. S. (1987) Free-form solid modelling with trimmed surface patches, *IEEE Computer Graphics & Applications* **7** (1) 33–43.

Chiyokura, H. (1988) *Solid modelling with DESIGNBASE*, Addison Wesley.

Dunnington, D. R., Saia, A., de Pennington, A. & Smith, G. L. (1989) Constructive solid geometry with sculptured primitives using ISOS techniques. In: W. Strasser (ed.) *Theory and practice of geometric modelling*, Springer-Verlag.

Farouki, R. T. (1987) Trimmed-surface algorithms for the evaluation and interrogation of solid boundary representations, *IBM J. Research & Development* **31** (3) 314–334.

Faux, I. D. & Pratt, M. J. (1979) *Computational geometry for design and manufacture*, Ellis Horwood, Chichester.

Foley, J. D. & van Dam, A. (1982) *Fundamentals of interactive computer graphics*, Addison-Wesley.

Goldman, R. N. (1983) Two approaches to a computer model for quadric surfaces, *IEEE Computer Graphics & Applications* **3** (6) 21–24.

Goult, R. J. (1989) Parametric curve and surface approximation. In: D. C. Handscomb (ed.), *The Mathematics of surfaces III*, Oxford University Press.

Hook, D. & Tiller, W. (1989) Boolean operations on 3D objects defined as collections of trimmed surfaces. In: W. Stresser (ed.) *Theory and practice of geometric modelling*, Springer-Verlag.

Lachance, M. A. (1988) Chebyshev economisation for parametric surfaces, *Computer Aided Geometric Design* **5** (3) 195–208.

Levin, J. Z. (1976) A parametric algorithm for drawing pictures of objects bounded by quadric surfaces, *Comm. ACM* **19** (10) 555–563.

Mäntylä, M. (1988) *An introduction to solid modeling*, Computer Science Press, Rockville, MD.

Nasri, A. H. (1984) Polyhedral subdivision methods for free-form surfaces, PhD thesis, School of Computing Studies & Accountancy, University of East Anglia.

Nutbourne, A. W. & Martin, R. R. (1988) *Differential geometry applied to curve and surface design*, Ellis Horwood, Chichester.

Pratt, M. J. (1985) Automatic blending in solid modelling — an approach based on the use of parametric geometry. In: *Proc. 4th UK/Hungarian Seminar on Computer Aided Design, Budapest, Oct. 1985*; Computer and Automation Institute of the Hungarian Academy of Sciences.

Pratt, M. J. (1989a) Applications of cyclide surfaces in geometric modelling. In: D. C. Handscomb (ed.) *The mathematics of surfaces III*, Oxford University Press.

Pratt, M. J. (1989b) Cyclide blending in solid modelling. In: W. Strasser (ed.) *Theory and practice of geometric modelling*, Springer-Verlag.

Pratt, M. J. (1990) Geometry for computer aided design. In: L. Piegl (ed.) *Key work in geometric modelling*, Butterworths.

Pratt, M. J. & Geisow, A. D. (1986) Surface/surface intersection problems. In: J. A. Gregory (ed.) *The mathematics of surfaces*, Oxford University Press.

Rockwood, A. P. & Owen, J. C. (1987) Blending surfaces in solid modelling. In: G. Farin (ed.) *Geometric modelling: algorithms and new trends*, SIAM, Philadelphia.

Rooney, J. & Steadman, P. (eds) (1985) *Principles of computer aided design*, Pitman/Open University.

Sederberg, T. W., Anderson, D. C. & Goldman, R. N. (1984) Implicit representation of parametric curves and surfaces, *Computer Vision, Graphics & Image Processing* **28** 72–84.

Várady, T. (1985) *Integration of free-form surfaces into a volumetric modeller*, Report 171/1985, Computer & Automation Institute of the Hungarian Academy of Sciences.

Várady, T. & Pratt, M. J. (1984) Design techniques for the definition of solid objects with free-form geometry, *Computer Aided Geometric Design* **1** (3) 207–225.

Várady, T., Martin, R. R. & Vida, J. (1989a) Topological considerations in blending boundary representation solid models. In: W. Strasser (ed.) *Theory and practice of geometric modelling*, Springer-Verlag.

Várady, T., Vida, J. & Martin, R. R. (1989b) Parametric blending in a boundary representation solid modeller. In: D. C. Handscomb (ed.) *The mathematics of surfaces III*, Oxford University Press.

Veenman, P. R. (1982) The design of sculptured surfaces using recursive subdivision techniques. In: *Proc. Conf. on CAD/CAM Technology in Mechanical Engineering, Massachusetts Institute of Technology, March 1982*; MIT Press.

Woodwark, J. R. (1986) *Computing shape*, Butterworths.

Woodwark, J. R. (1987) Blends in geometric modelling. In: R. R. Martin (ed.) *The mathematics of surfaces II*, Oxford University Press.

10

Modelling of free-form surfaces using GENSURF

Don Catley
Manager Software Engineering, BMT CORTEC Ltd, UK

The chapter reviews certain aspects of the mathematical background to surface definition and emphasizes the benefits of *B*-spline surfaces, particularly when used for manipulating a free-form design. The generally applicable surface definition, fitting and manipulation system designated GENSURF, is described. This system combines several desirable attributes such as user-friendliness, flexibility, and robustness with a sound mathematical background on surface modelling.

When using GENSURF, geometric and other primitives can be generated and transformed into appropriate patch data files, which are then fitted to defined regions of the surface. More general surface regions can be defined based on 'tracked' data and fitted in a similar manner. Manipulations of the surface are carried out by consideration of the displacements required to appropriate polygon points, the control vertices which define the surface. The patches in GENSURF are based on a bicubic, *B*-spline formulation, and are essentially four-sided. However, degenerate situations can be handled satisfactorily. Furthermore, sets of surface curves can be defined.

GENSURF is applicable to the design and definition of any free-form surface and to any industry where free-form surfaces are an integral part of the design. Routines are available for the calculation, interrogation, and manipulation of the *B*-spline curve and patch data for uniform and non-uniform splines. These routines, which are written in FORTRAN-77, are stored in an object library which can be linked to the user's own program.

A customized version of GENSURF, designated HULLSURF, has successfully been applied to a variety of ship forms, SWATHS, submarines, and propeller geometries. Example applications are given in the paper which highlight the salient features of the system. These include patch definition, patch boundary continuity requirements, bending, filleting, surface-to-surface intersections with planes or other general surfaces, and patch clipping.

INTRODUCTION

One of the fundamental and central processes in design is the representation of a three-dimensional shape in term of two-dimensional drawings. For complex geometrical shapes the most widely used representation is via sectional plans, and three mutually orthogonal views are constructed. In the shipbuilding and other industries these views are well known and are collectively called the lines plan.

Over the past twelve or so years BMT (formerly BSRA) have developed an

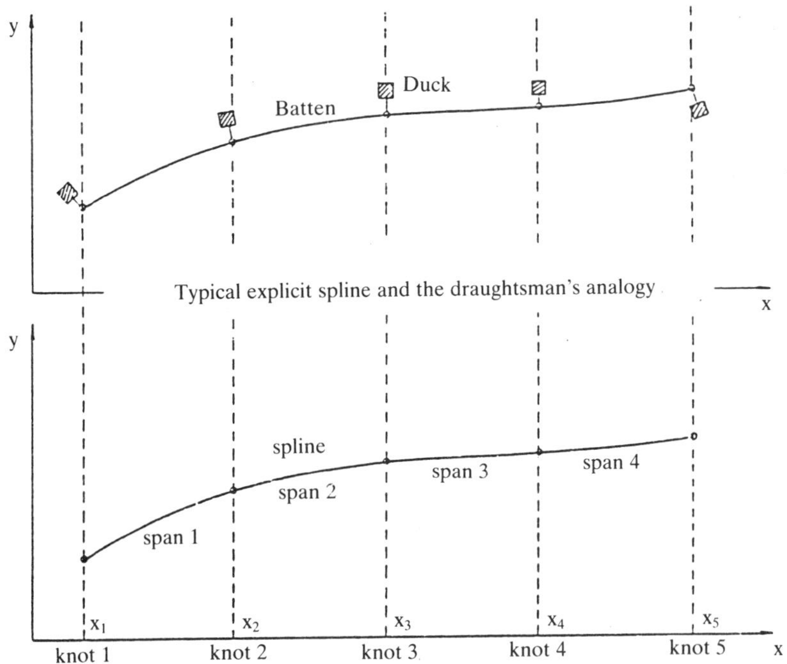

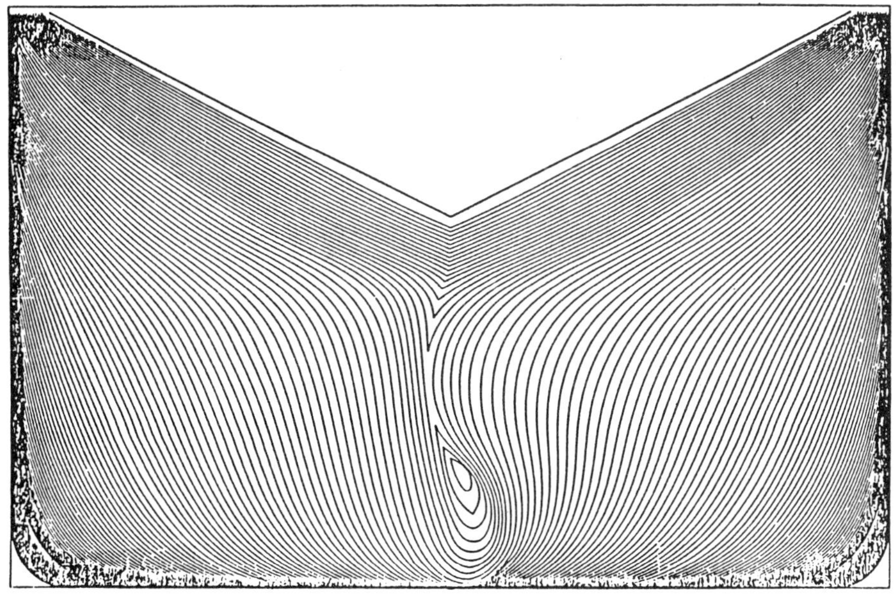

B-LINES sections for an asymmetric stern

Fig. 1 — Modelling traditional lines.

efficient computerized lines system, designated B-LINES, which is based on cubic *B*-spline curves. The name spline originates from the draughtsman's batten, although it has now developed into a broad subject in the field of mathematics. Fig. 1 shows how a *B*-spline curve models the traditional practice of a loftsman and illustrates the B-LINES system applied to an asymmetric form. Customers have now been weaned onto the use of computer-aided design techniques, and their acceptance and enthusiastic use of B-LINES has played a major part in this transformation.

Fairing of the sectional plans does not imply fairness in another view, thus traditional fairing is an iterative process in any surface manipulations. This process requires the patience of an experienced loftsman. B-LINES has drastically reduced the time needed for fairing and has been used successfully as a design tool. For practical, as well as for aesthetic, reasons, it is prefereable to define each point of the surface uniquely rather than by complementary line sets. In particular, there are requirements for an accurate representation of curvature and other surface features. BMT have developed the HULLSURF system which is based on a patch definition; typically 30 patches define the geometry of a hull. The *B*-spline curve theory has been extended in a natural way to define the surface, using a concise mathematical model.

Fig. 2 illustrates some manipulations of a typical surface and curvatures used to quantify the fairness. HULLSURF offers the means whereby the surface implied by section lines can be defined and controlled efficiently throughout any design changes which are imposed.

With availability also of the BMT PATCHGEN system, geometric primitives such as special curves, portions of spheres, cylinders and axisymmetric bodies, etc., can be generated and manipulated in a semi-automatic manner. Fig. 3 illustrates many of the features of the software. In one day the previously conventional forebody of this ship was re-fitted with a bulbous bow, and a dramatic improvement in flow characteristics was obtained. By using HULLSURF the designer is able to expediently assess alternative forms and the effect of design changes; data are supplied through to the production process.

A considerable amount of development has now been carried out at BMT to utilize the advantages of *B*-spline techniques for both curve and surface fitting and manipulation. The total surface definition software has now been consolidated into a single module, designated GENSURF, which is applicable to a broad range of industry. Salient features include blending, filleting, offsetting, general surface-to-surface intersections, patch clipping, and surface interrogation for geometrical properties such as areas, volumes, and curvatures, including the drawing of principal curvature tufts. The mathematical model is completely general; it allows for the definition of a large variety of free-form surfaces so that unconventional designs can be modelled. Figs 4 and 5 show GENSURF applied in a straightforward manner to the benchmark tests of Ball (1984) and Munchmeyer (1987) [1] and [2]. Fig. 6 is a collage of various other applications of the software.

Other BMT software modules are able to exploit to advantage the surface definition which is available from GENSURF. Furthermore, the software has become an integral part of major systems used by customers, for example, the US Naval Sea Systems Command (NAVSEA) — the agency responsible for the design of naval ships. Other customers include companies from the UK, Canada, South America, Scandinavia, Germany, and the Far East.

The software is based on the in-built, interactive graphics facility INGOTS and a simple to use, single-level command structure. It is user-friendly and is portable to a wide range of mainframe and microcomputers. British and US Naval Standards of approval have been obtained.

MATHEMATICAL BASIS OF GENSURF

Ideas developed for *B*-spline curves by Schoenberg (1946), Reisenfeld (1978), Cox (1972), and De Boor (1972) are well established. The extension of the techniques to three-dimensional surfaces or even surfaces of higher dimension is relatively straightforward. Faux & Pratt (1979) provide good general reading.

With the advent of computerized technology, several surface definition schemes are now available; see, for example, the conference proceedings of Catley *et al.* (1985a). These tend to be based on Coons patches, Bézier patches, or one of various types of spline surface. Various methods have been briefly reviewed by Catley *et al.* (1985b) and Catley *et al.* (1984, 1985b), Clark (1981), Catley (1987), and Applegarth *et al.* (1988) have outlined the approach adopted at BMT which offers a formulation most suitable for the interactive design, manipulation, and interrogation of surfaces. The software is applicable to any rolled or sculptured surface.

Although several methods have been proposed, the most usual, and the one adopted in GENSURF, is the tensor product surface defined by using *B*-spline basis functions $N_{i,k}$ with appropriate weighting (Catley *et al.* 1985b):

$$x(s,t) = \sum_{i=1}^{n} \sum_{j=1}^{m} X_{i,j} N_{i,k}(s) N_{j,k}(t)$$

$$y(s,t) = \sum_{i=1}^{n} \sum_{j=1}^{m} Y_{i,j} N_{i,k}(t) N_{j,k}(t)$$

$$z(s,t) = \sum_{i=1}^{n} \sum_{j=1}^{m} Z_{i,j} N_{i,k}(s) N_{j,k}(t)$$

where two parameters s and t are required. In the adopted approach the order of the basis functions is $k=4$, that is, the surface is locally a bicubic, *B*-spline surface.

These equations form a mapping from two-dimensional space (s,t) to the real three-dimensional space (x,y,z). The surface is subdivided by boundary curves into regions referred to as surface patches. The coefficients $X_{i,j}$, $Y_{i,j}$, and $Z_{i,j}$ are the polygon point vertices which define the surface. Based on the data points for the patch, a two-way spline fit is performed to defined the vertices. The data points are usually interpolated, but, by using a weighted fit, regions of the surface where the data points are not fair can be approximated.

In the mathematical model, the hull surface is divided into surface patches which are based on sets of defining curves lying on the surface. The defining curves are

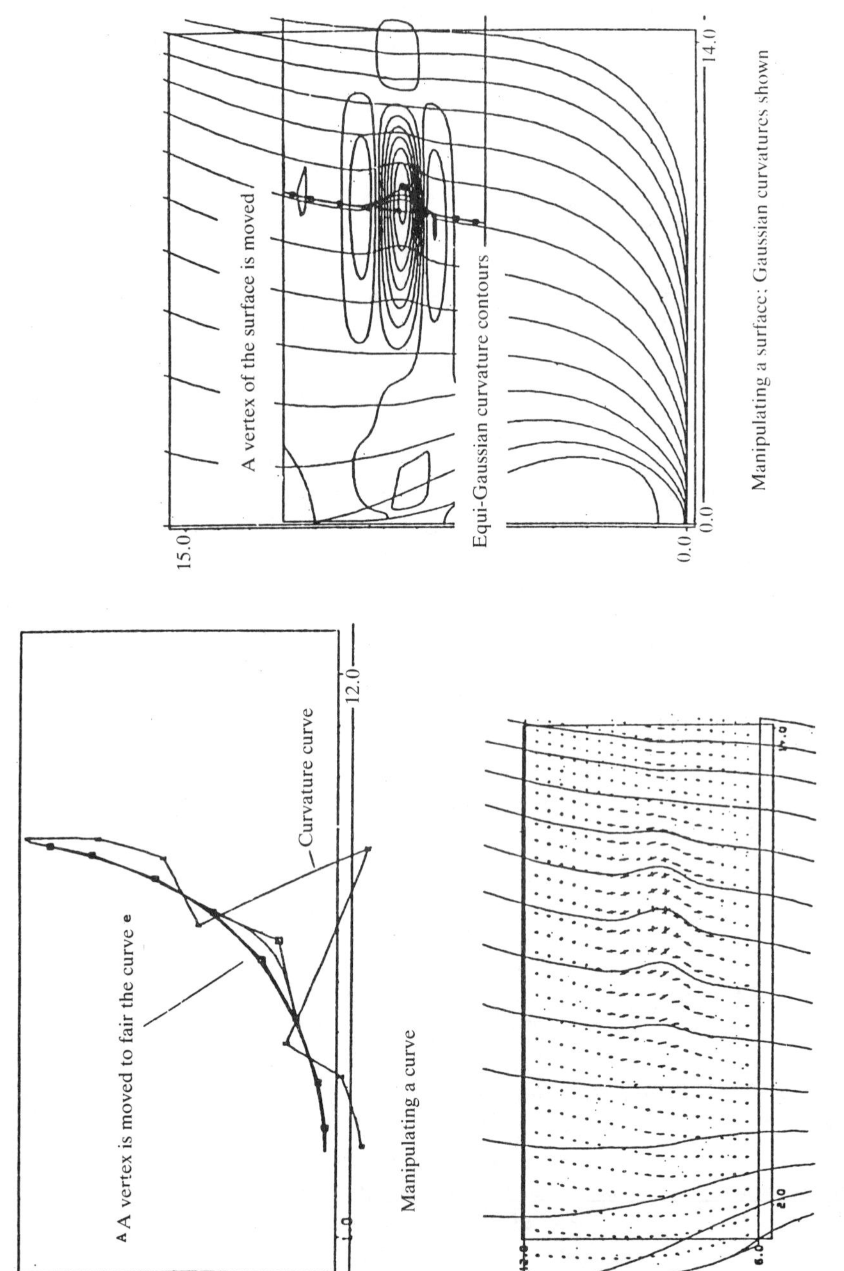

Fig. 2 — Typical curve and surface manipulations.

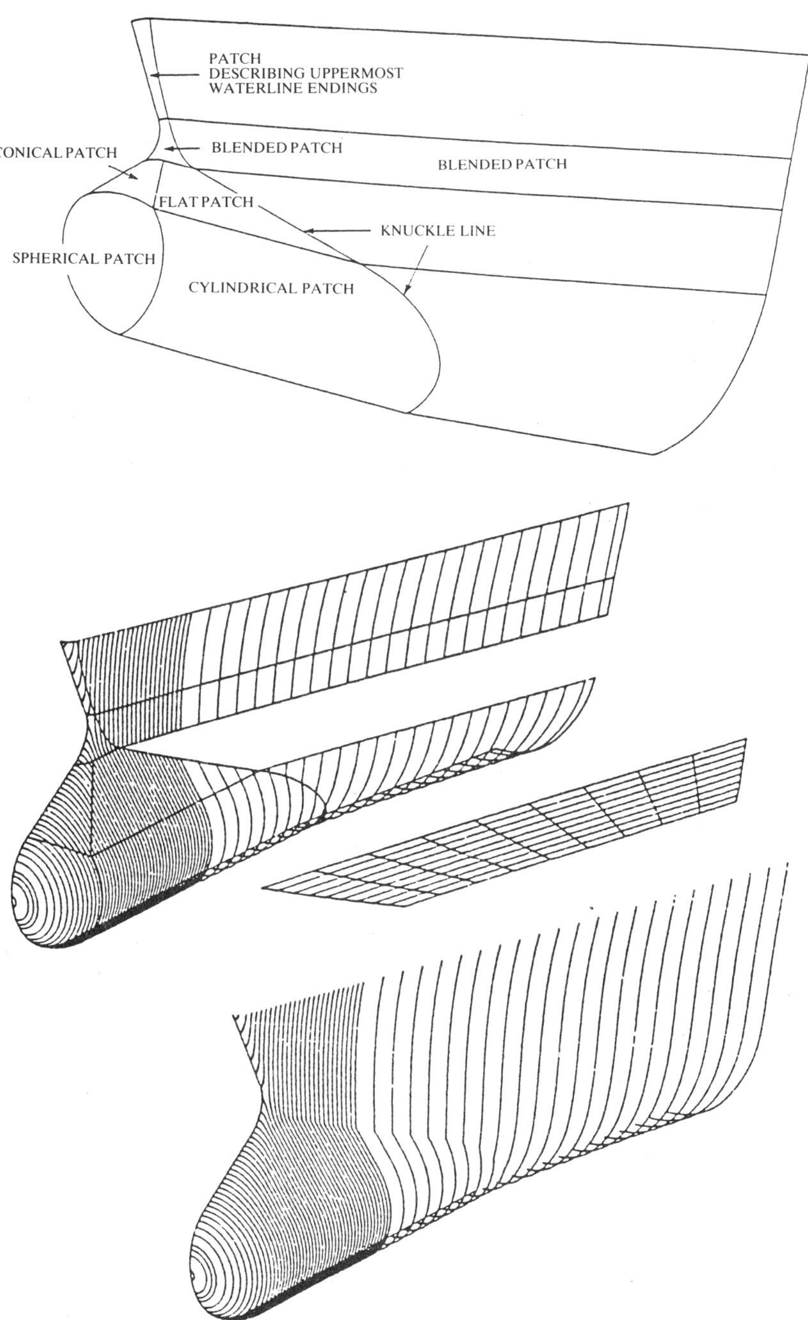

Fig. 3 — Fitting a bulbous bow to improve flow characteristics.

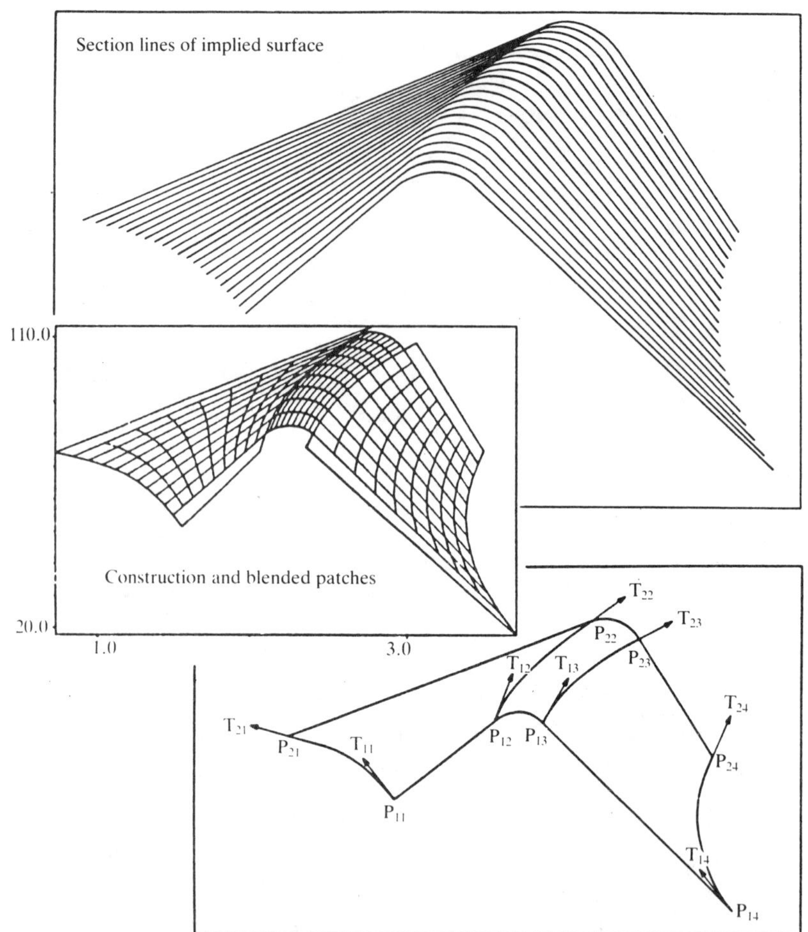

For the first patch

Form	Data generated for benchmark		
4			
	75.000	55.0000	0.0000
	76.000	53.8000	− 10.0000
	77.000	52.6000	− 20.0000
	78.000	51.4000	− 30.0000
4			
	105.000	80.0000	0.0000
	104.400	78.2000	− 10.0000
	103.800	76.4000	− 20.0000
	103.200	74.6000	− 30.0000

For the third patch

Form	Data generated for benchmark		
4			
	42.0000	73.0000	250.000
	40.1000	73.4000	260.000
	38.2000	73.8000	270.000
	36.3000	74.2000	280.000
4			
	135.0000	111.0000	250.000
	136.3000	111.8000	260.000
	137.6000	112.6000	270.000
	139.9000	113.4000	280.000

For the first patch

Form	Data generated for benchmark		
4			
	120.000	80.0000	0.0000
	119.300	78.6000	− 10.0000
	118.600	77.2000	− 20.0000
	117.900	75.8000	− 30.0000
4			
	180.000	25.0000	0.0000
	181.800	22.6000	− 10.0000
	183.600	20.2000	− 20.0000
	185.400	17.8000	− 30.0000

For the third patch

Form	Data generated for benchmark		
4			
	149.0000	106.0000	250.000
	150.3000	106.7000	260.000
	151.6000	107.4000	270.000
	152.9000	108.1000	280.000
4			
	173.0000	71.0000	250.000
	173.6000	72.3000	260.000
	174.2000	73.6000	270.000
	174.8000	74.9000	280.000

Fig. 4 — GENSURF applied to the 'Loughborough benchmark' using blending.

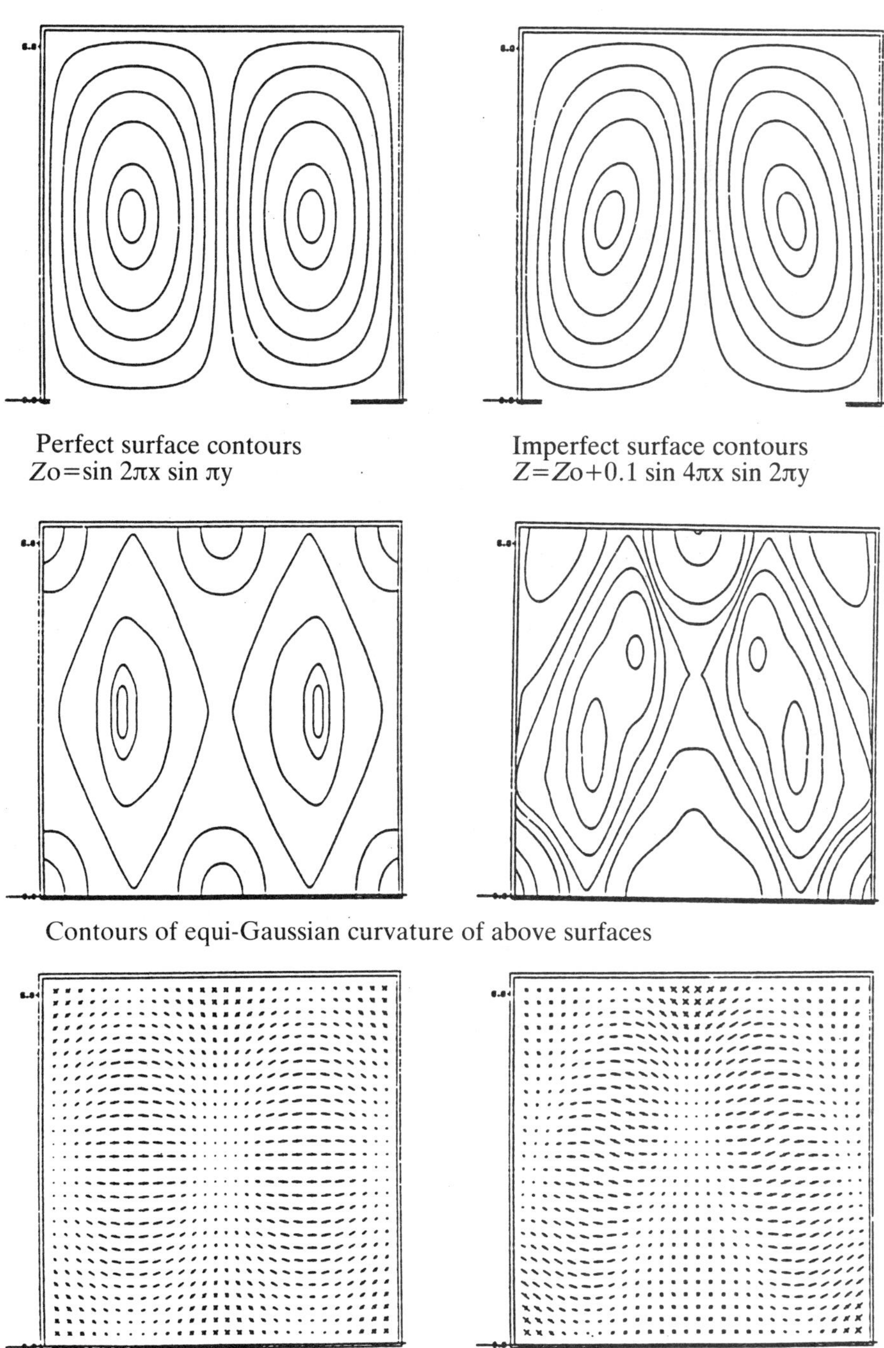

Fig. 5 — GENSURF applied to an imperfect surface benchmark.

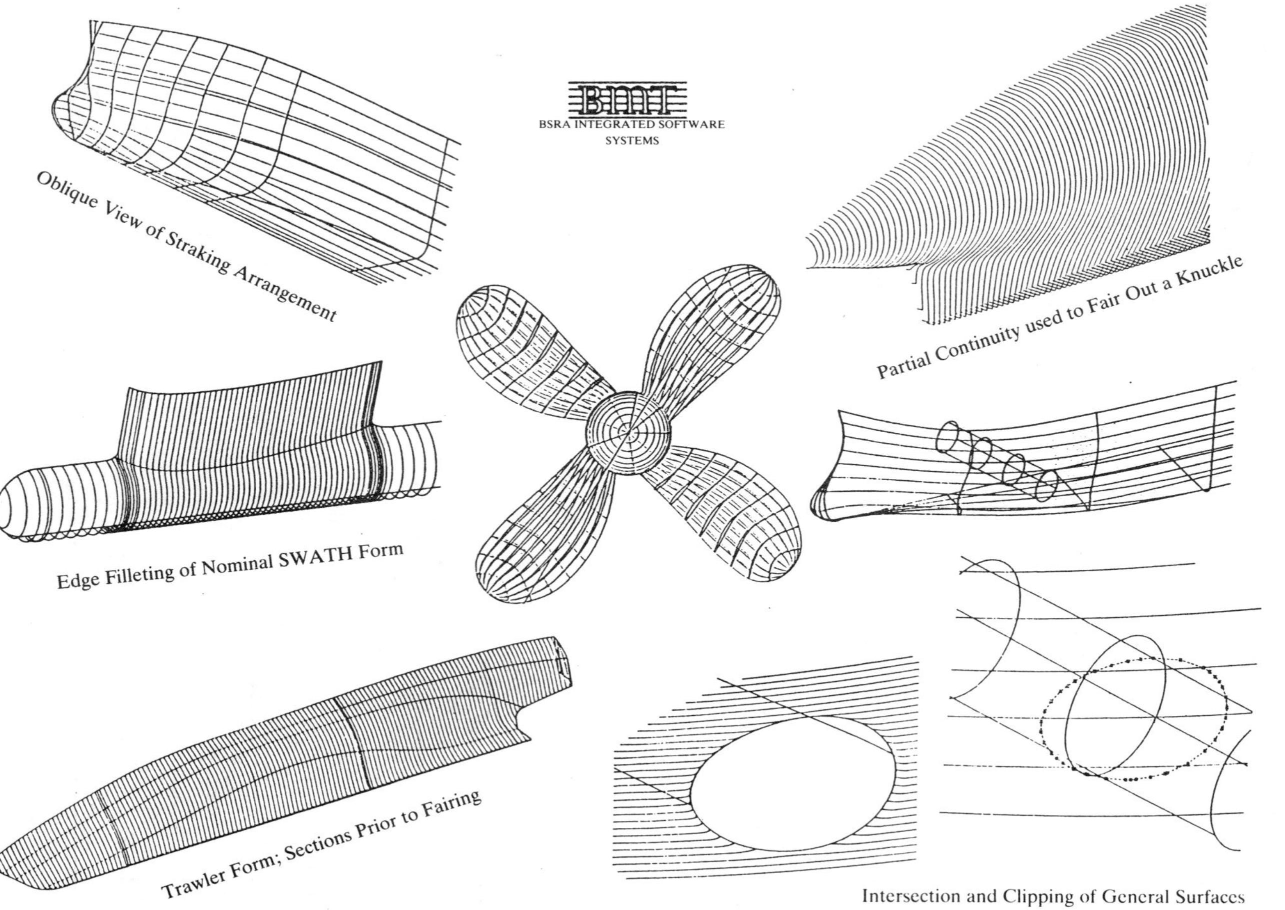

Fig. 6 — Example surfaces designed and fitted using GENSURF.

themselves specified by sets of data points. The polygon points are used to control tangency continuity, where required, at the patch boundaries.

There is no effective limit on the number of patches that may be defined, and there is a greater deal of flexibility in the allowed data input procedures. The usual restriction of a rectangular array of regularly spaced points, associated with acceptable *B*-spline surface interpolation, has been overcome. For each patch, two independent knot sets correspond to a specified 'net size' are required to define the necessary basis functions, and these form a rectangular grid for a surface patch. Additional coincident knots are included at the edges of the patch. Shape changes to associated *B*-spline curves and surface regions can be achieved by manipulating the appropriate defining polygon points.

The definition may be interrogated whenever any information is required, for example surface information required for the hydrodynamic analysis of the form, the shape of a frame, displacement at a particular depth, the coordinates of a point, etc. Volumes, areas, curvature, and distance functions can also be calculated. Various applications of the software have been given (Catley 1986, 1988, 1989, Catley & Romero 1987, Catley & Whittle 1987, Catley & Marshall 1989, Vardon 1986).

THE GENSURF ROUTINES

The GENSURF mathematical library is comprised of routines for the calculation, interrogation, and manipulation of *B*-spline curve and patch data for uniform and non-uniform splines. For convenience several utility routines are also included. The routines, which are written in FORTRAN 77, are stored in an object library which can be linked to the user's own program. B-LINES and HULLSURF are examples of such customized software.

The routines in this library may be considered as forming two sets, i.e. for *B*-spline curves and for *B*-spline surfaces. Within these sets are various usage groups as follows:

- *B*-spline curves
  - Curve fitting
  - Curve interrogation
  - Curve manipulation
  - Curve output, vectors and polygon points
- *B*-spine surfaces
  - Surface fitting
  - Surface generation
  - Surface interrogation
  - Surface manipulation
  - Surface output; vectors and polygon points

Curve fitting

In GENSURF each point of a data curve is assigned a 'point type'.

- 0 (zero), the spline has curvature continuity at this point. This is known as an 'ordinary' point.

- 1, the spline has discontinuous curvature but continuous tangent at this point. That is to say, that the second directional derivative is discontinuous. This fit will give a straight line from the point to the next data point. This is known as a 'tangent' point.
- 2, the spline has discontinuous tangent (and curvature) at this point. This is known as a 'knuckle' point.

There are a variety of curve fit routines. Optionally, the knot set may be specified rather than generated by GENSURF. Tangency control may be imposed at one or more points. This may be 'automatic', in which case the fit will enforce constant tangency between two specified points, or 'manual', in which case tangency values are specified by the user. The option is effected by a suitable knot set and associated polygon points. The B-LINES system uses these routines to advantage. The control which is possible for such parametric splines is illustrated in Fig. 7.

Specific routines are available for fitting two-dimensional geometric primitives, viz: arc, parabolic segment, hyperbolic segment, line segment, NACA, and EPH (customized) profiles.

Curve interrogation

Routines may be called:

- To calculate the area between specified points or parameter values on a spline.
- To calculate the derivatives of a given *B*-spline curve at specified values of the *X*, *Y*, and *Z* coordinates.
- To calculate the girth between specified points or parameter values on a given *B*-spline curve.
- To interpolate a point on a spline at specified values of given *X*, *Y*, or *Z* values or parameter values on a given *B*-spline curve.
- To calculate first or second moments between specified points on a given *B*-spline curve.
- To calculate the coordinates of points lying at a specified girthwise distribution along a given *B*-spline.
- To obtain the coordinates of the nearest points of the two-*B*-spline curves and their distance apart.
- To intersect a *B*-spline curve with a general plane, or surface patch.

Example applications of these routines are the BMT systems:

B-LINES — fairing of lines and hydrostatic analysis.
MINT — topological definition of major internal structure.
HULLDAS — hull design analysis.

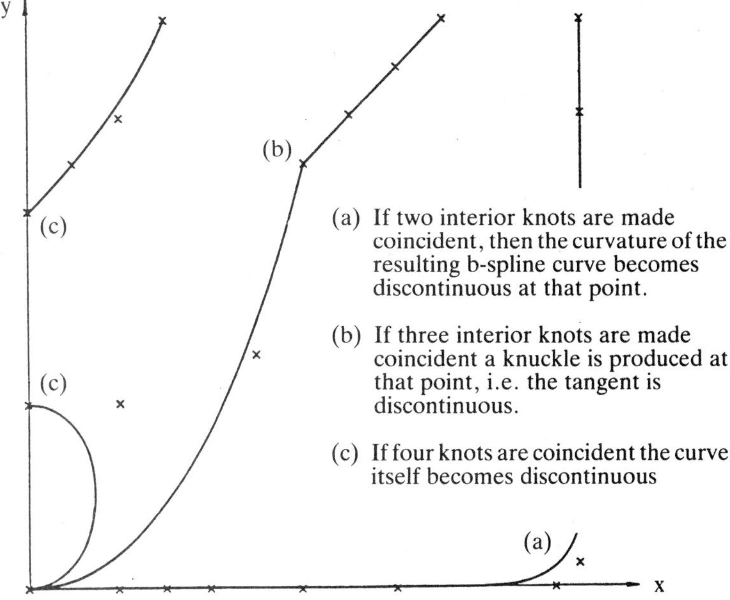

(a) Discontinuities due to multiple knots

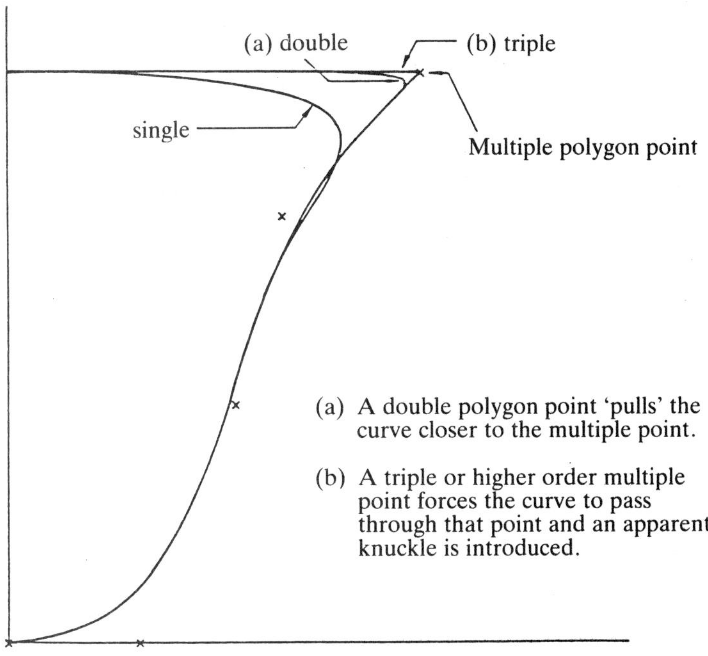

(b) Control via multiple polygon points

Fig. 7 — Effect of knot selection on a curve.

Curve manipulation and output

The following routines are available:

- To scale, rotate, translate or shear a given *B*-spline curve.
- To clip a *B*-spline curve to lie between two specified parameter values.
- To insert a polygon point into a *B*-spline curve.
- To remove a polygon point from a *B*-spline curve.
- To calculate and return the number and coordinate values of vectors needed to draw a smooth curve through a given *B*-spline curve for a given tolerance.

Surface fitting and generation

A rectangular grid of data points which may be either regular or irregular may be fitted to give a bicubic *B*-spline surface patch. Optionally knot values can be specified. This represents a minimum of constraint for the input data which must only be based on 'tracked' data, i.e. sets of ordered points on non-crossing curves.

Specific routines are available for modelling three-dimensional geometric primitives. These form the basis for surface definition in the HULLSURF and MINT systems.

- Specification of two boundaries with an interior straight line fit to give a FORMED patch.
- Specification of two conic boundaries to fit a DEVELOPABLE patch.
- Specification of one boundary and a vector to fit a RULED patch.
- Fit defining a unit PLANE, CYLINDER, SPHERE, or CONE.
- Fit to one boundary curve rotated about a given axis to give an AXISYMMETRIC patch.
- Fit to two given boundaries with a specified interior swept curve fit.
- Fit to give tangency continuity at boundaries of the patch with up to four specified surface patches.
- Fit directly to a set of curves each defined by the same number of polygon points.

A routine is also available for the blending of a bicubic *B*-spline surface patch between two specified patches defined by their polygon points. See for example, Figs 3 and 4.

Surface interrogation

Routines may be called:

- To obtain the coordinates and reference numbers of the common corner points of two adjacent *B*-spline surface patches.
- To calculate the surface area of one or more given *B*-spline surface patches.
- To calculate the volume enclosed by one or more given *B*-spline surface patches.
- To interpolate a point on a given *B*-spline surface patch, i.e. given *X*, *Y*, find *Z*, etc.
- To calculate the *X*, *Y*, *Z* coordinates defining a planar intersection of a given *B*-spline surface patch at a specified *X*, *Y*, or *Z* position.

- To intersect a given B-spline surface patch with a specified general plane and return the X, Y, Z coordinates of the segments of the intersection curve.
- To intersect two given B-spline patches.
- To interpolate a grid of intersection points on a given surface..
- To calculate points lying on the given B-spline surface patch at parametric mesh intersection points.
- A check of the mean and variance of the fitted surface deviation from the given data points.
- To calculate the minimum distance of a given X, Y, Z position from a given B-spline surface patch.
- To calculate the Gaussian curvature at interior parametric positions to a given B-spline surface patch and return the Gaussian curvature values, with a B-spline fit to Gaussian curvature values and polygon points returned.
- To calculate the directions and magnitudes of principal curvature at interior parametric positions to a given B-spline surface patch. Optionally to interpret the principal curvature values of a given B-spline surface patch as tufts on the surface.
- To integrate a scalar function, for example, pressure, defined at nominal centroid locations of a surface patch which is defined with reference to the facet corner points.

As an illustration of the latter of these routines, tests are carried out for a sphere and a horizontal cylinder, each half immersed in a nominal fluid, with the pressure (i) uniform and (ii) varying linearly with depth. A comparison with the analytical results is given in Table 1.

Table 1 — Comparison of GENSURF and analytical results

Body	Pressure function	Number of facets	Force calculated	
			GENSURF	Analytical
Sphere	Uniform	14 × 14	314.1	314.2
Sphere	Varying as depth	14 × 14	2094.6	2094.4
Cylinder	Uniform	8 × 3	60.0	60.0
Cylinder	Varying as depth	8 × 3	471.7	471.2

Accuracy can be improved by increasing the number of facets.

Surface manipulation and output

The following functions are available:

- To scale, rotate, translate or shear a given B-spline surface patch.

- To fillet the common boundary of two *B*-spline surface patches; tangent or curvature continuous fillet.
- To clip a *B*-spline surface patch to lie between two specified pairs of parametric values..
- To intersect a given *B*-spline surface patch at a given *X*, *Y*, or *Z* value or general plane or *B*-spline surface patch and return the *X*, *Y*, *Z* coordinates of vectors or polygon points to represent the segments of the intersection curve for a given tolerance.
- To return *X*, *Y*, *Z* coordinates of vectors to represent a mesh of parametric curves (facet points) on a given *B*-spline surface patch for a given tolerance and scale.
- To return *X*, *Y*, *Z* coordinates of vectors to represent the boundary curves of a given *B*-spline surface patch for a given tolerance and scale.

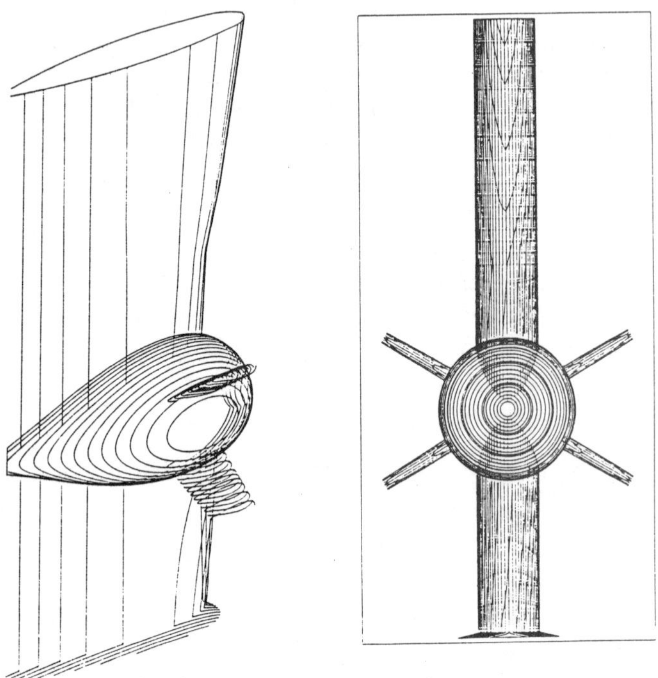

Fig. 8 — Sections of a rudder.

These routines are used, for example, by the HULLSURF system for interactive surface design and analysis. As illustrations, Fig. 8 shows section curves for an example rudder and Fig. 9 shows facet points for a hull and SWATH surfaces.

GENERAL COMMENT

The GENSURF system has been designed to allow the user a large degree of freedom in the sequence in which the surface items are created or manipulated to

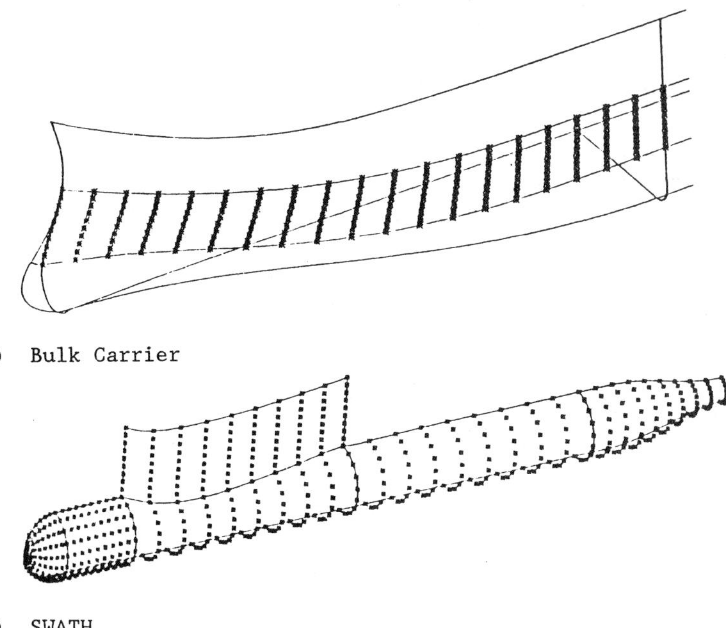

Fig. 9 — Facet points of example surfaces.

define a particular form. The routines have been structured so as to suggest a sequence for creating and modifying the surface. The approach is valid for most forms but can be varied as appropriate by the user.

The ease with which particular features can be incorporated or removed from a design depends on the need, if any, to reorganize the patch boundaries. However, the patching can be exploited to advantage in controlling the definition of the surface. Usually, certain regions will be critical to the overall design of the surface in terms perhaps of hydrodynamic or aerodynamic criteria or else production difficulty and cost.

The industries which could find benefit from the use of a tool such as GENSURF are numerous, but include:

- aerospace
- car design and production
- nuclear industry
- petrochemicals
- glass
- broad aspects of the marine industry
- propeller design and manufacture
- educational establishments
- domestic appliances
- sculptured surfaces

- shoes
- data for solid modelling involving complex surfaces
- contouring by the fitting of scattered data

ACKNOWLEDGEMENTS

The author would like to thank the Board of BMT CORTEC Ltd for permission to publish this paper. As the work described has been progressed over several years, acknowledgement is due to a number of colleagues at BMT both past and present.

REFERENCES

Applegarth, I., Catley, D. & Bradley, I. (1988) Clipping of *B*-spline patches at surface curves, *IMA Conference on the Mathematics of Surfaces III*, ed. D. C. Hanscomb, Oxford, 1988, to appear.

Ball, A. A. (1984) *The Loughborough Benchmark*, CADCAM International, October.

Catley, D. (1986) Towards in integrated procedure for the mathematical modelling of ship vibrations: *POLYMODEL* 9 *Conference on Industrial Vibration Modelling, BMT, Wallsend, May* 1986: Publisher Martinus Nijhoff.

Catley, D. (1987) Applications of the general surface definition and manipulation system, GENSURF, *IMA Conference on the Mathematics of Surfaces II*, ed. R. R. Martin, Clarendon Press.

Catley, D. (1988) Illustrated applications of the GENSURF system, *CADMO* 88, Southampton, 1988.

Catley, D. (1989) Interactive definition of shell plating and its production, *PRADS* '89, *Bulgarian Ship Hydrodynamics Centre, Varna*, 1989.

Catley, D. & Marshall, G. (1989) Software for ship design through to production using mainframes, workstations or PC's IRPEM (CNR) *Symposium Technics and Technology in Fishing Vessels, Ancona, Italy, May* 1989.

Catley, D. & Romero, J. L. (1987) BMT hull design software, *STG Sprechtag CAD/CAM in der Schiffstechnik, Hamburg March 1987.*

Catley, D. & Whittle, C. (1987) Exploitation of an advanced surface definition capability for hull form design and analysis, *PRADS 87 Trondheim, Norway, June 1987.*

Catley, D., Okan, M. B. & Whittle, C. (1984) *Unique mathematical definition of a hull surface, its manipulation and interrogation*: WEMT, Paris, July 1984.

Catley, D., Whittle, C., & Thornton, P. (1985a) Applications of an advanced hull surface definition system in ship design: *ICCAS 85*, Trieste, Italy.

Catley, D., Davidson, G. H., Okan, M. B., Whittle, C. & Thornton, P. (1985b) A system for general surface definition and manipulation, *NATO Advanced Study Institute Conference on Fundamental Algorithms for Computer Graphics, Ilkley, Yorkshire*, April 1985.

Clark, A. P. (1981) *Spline methods for Computer Aided Definition of Ship's Lines*: BSRA SRD Report, Feb.

Cox, M. G. (1972) The numerical evaluation of *B*-splines: *J. Inst. Maths. Applics.*, **10**.

De Boor, C. (1972) On calculating with *B*-splines: *J. Approx. Th.*, **6**.

Faux, I. D. & Pratt, M. J. (1979) *Computational geometry for design and manufacture*, Ellis Horwood.

Munchmeyer, F. (1987) On surface imperfections, *IMA Conference on the Mathematics of Surfaces II*, ed. R. R. Martin, Clarendon Press.

Reisenfeld, R. F. (1978) *Recommendations for Computer Utilization in Shipbuilding*: University of Utah Technical Report (companion report to Chen, R. and Cuthill, E., 1978).

Schoenberg, I. J. (1946) Contributions to the problem of approximation of equidistant data by analytic functions: *Quart. Appl. Math.* **4**.

Vardon, J. R. (1986) CAD in shipbuilding (ship hull design and structures); *Tutorials for CADMO 86*, Washington, USA, Sept. 1986.

11

The geometry of flexible membranes in wave and hydro energy devices

F. P. Lockett, L. J. Duckers, B. W. Loughridge, A. M. Peatfield, M. J. West, and **P. R. S. White**
Energy Systems Group, Coventry Polytechnic, UK

INTRODUCTION

In the early 1970s the surge in world oil prices caused a flowering of interest in renewable energy sources. In particular, wave energy, with its huge resource potential, was regarded as very promising, and ten research groups in Britain produced costed designs for a 2-GW power station off the Outer Hebrides. Assessment of these schemes, based on a variety of devices, showed most to be uneconomic. Those which were judged to be most cost effective were the DUCK, invented by Professor Stephen Salter at Edinburgh University, and the CLAM, pioneered by the Energy Systems Group at Coventry Polytechnic. The central problem of wave energy is to harness huge forces at low velocity, in contrast to the low forces at high velocity more typical of normal machinery. The DUCK device seeks to solve the problem with clever mechanical and hydraulic engineering achieving high efficiency but at high and possibly prohibitive cost. On the other hand, the CLAM aims at constructional simplicity to get good performance at low cost. It employs a pneumatic power take-off system, making the waves drive air turbines coupled to generators, using flexible rubber membranes as the interface between the sea and the air system to convert wave to air power.

THE CLAM

Almost all of the operational wave power devices built so far around the world have been of the oscillating water column design, either as buoys or shore-mounted. These, like the CLAM, use the Wells air turbine which, having blades of symmetrical aerofoil cross-section, rotates in the same sense whichever the direction of the air flow, allowing it to be driven by the reversing air flow induced by wave motion. The

CLAM device currently envisaged is a rigid floating torus some 60 m in diameter consisting essentially of 12 interconnected air cells with rectangular membranes on their outer faces separating the sea from the closed air system.

Figs 1 and 2 show a $\frac{1}{15}$th scale model before launch and during tests on Loch Ness.

Fig. 1 — Scale model of CLAM, before launch.

Fig. 2 — Scale model of CLAM, during tests.

The air system is partly inflated, with the membranes approximately $\frac{2}{3}$ submerged, then differential wave action around the torus causes air to be pumped back and forth between the cells. This flow is resisted by Wells turbines, one per cell, which power the generators. With compliant mooring and low freeboard, the device experiences moderate mooring and structural forces, even in storm conditions. With design development, the membrane and power take-off equipment could be produced simply and cheaply. These factors indicate the potential for economic volume production and competitive energy cost. In 1985 this device was assessed bv independent consultants as being capable of producing an annual average power output of 600 kW at a cost of 6 p/kWh at today's prices. Regrettably, this show of promise was deemed to be insufficient to merit further goverment support, and no more development has been done since then. However, in the wake of current renewed interest in non-fossil energy sources, the Energy Systems Group in Coventry have hopes of future support for a resurrection of work culminating in the deployment of a full scale device at sea.

THE BORROWASH HYDRO SCHEME

In the meantime, since 1984, the Group has remained active with funding from RMC Group plc and the EEC to build a novel low-head hydroelectric station at an old mill site at Borrowash in Derbyshire, using the same technological ingredients as the CLAM.

The scheme at Borrowash consists of an inclined duct some 30 m long and 6 m wide linking collecting ponds at high and low water levels with a head difference of about 2.7 m. A flexible rubber membrane runs down the centre plane of the duct, its edges clamped to the sides of the duct and the floor at its ends, thereby separating water above from air below it. The membrane is slightly wider than the duct, which allows it to seal against the arched duct roof. A flexible septum connects the membrane to the duct floor approximately half way down the device, dividing the air space beneath the membrane into two chambers. These are connected externally to form a closed air system by a smaller duct containing the Wells turbine coupled to a generator. The arrangements are shown schematically in Fig. 3. If the system is half filled with air, a slug of water enters the duct, pushing air from the top to the bottom chamber via the turbine. Eventually nearly all of the air is displaced as the slug descends, and the lower chamber, now fully inflated, becomes pressurized, forcing air back into the top chamber which inflates to seal the duct entrance. The slug then continues to descend, displacing air from the bottom chamber and ultimately exits via the duct, allowing the cycle to recommence. The device is thus self-oscillating, inducing reversing air flow through the turbine.

MEMBRANE CONSTRUCTION

Design and construction of the membrane at Borrowash has been carried out by Avon Rubber plc working in close collaboration with Dr L. J. Duckers of the Energy Systems Group. The membrane material is a blend of natural and polybutadiene rubber reinforced with cross-ply cords, approximately 4 mm thick. An identical design is envisaged for the CLAM wave energy application. The material strength,

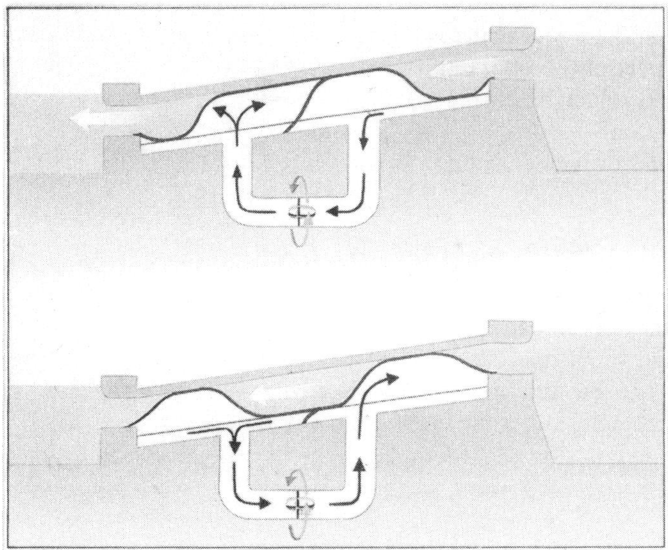

Fig. 3 — Borrowash hydro scheme.

and hence its geometry in use, is determined predominantly by the cord fibres; the rubber serves mainly to hold the closely packed net of cords in their relative positions and to seal the air from the water. At Borrowash the two plys of cords are aligned at angles of 15° to the transverse direction, across the duct. This provides the strength to support the 100 tonne slugs of water as they descend and the longitudinal stretch down the duct necessary to allow the membrane to deform and seal against the roof at either end of the slug. For the CLAM, most membrane strength is required in the vertical direction to oppose buoyancy forces, while resilience is needed horizontally to allow the membrane to flex between convex and concave positions without severe buckling during operation. This can be achieved again by cross-cording at 15° to the vertical, so each CLAM membrane panel would be roughly of the same dimensions as one half of the duct membrane at Borrowash.

MEMBRANE GEOMETRY

The objective of the geometric modelling for our applications is to compute the equilibrium shape taken up by a membrane with specified tensile (and to a far lesser extent flexural) properties, size, edge location, and given air pressure and water loads. This problem is akin to those met in the science of textile fabrics, and it seems that progress has been significant only in 2-dimensional or simple 3-dimensional cases. Hearle presents a useful survey in Amirbayat & Hearle (1985). The problem is very different from the more common and tractable type in which we seek a small deflection from a known position, for example in the stress analysis of a rigid structure. Indeed, on our applications, small changes in the applied forces can produce large changes in the membrane geometry, unlike the situation in the analysis of, for example, cable roof structures where such sensitivity would be disastrous. Our devices work pecisely because of this physical instability which is unfortunately often reflected in the reluctance of iterative numerical solutions to converge. For these reasons we can only report, at present, work based on 2-dimensional analysis which has nevertheless provided useful information for design and on performance.

CLAM MEMBRANE CROSS-SECTION

Consider a vertical cross-section l through the CLAM membrane. Since there is little horizontal variation in loads and significant horizontal compliance of the material, end effects are small, and so over most of its width the cross-sectional shape is uniform and can be determined by a 2-dimensional computation. The physical condition which determines the curve l is the relation

$$pR = T$$

where p = pressure difference across the membrane,
R = radius of curvature, and
T = tangential membrane tension per unit width.

Since the external forces on the membrane are all normal to it, the tension T is constant along the curve. We determine l by a sequence of points $Q_0, Q_1, \ldots, Q_N$, regularly spaced along the curve, using a shooting method to fit the boundary conditions. At its base the curve either runs out from the structure at some angle α, say, to the horizontal, or lies against the structure for some contact distance c, say. In a general position, the point Q_{k+1} is determined by using the condition that the points Q_{k-1}, Q_k, Q_{k+1} lie on a circle of radius $R = T/p_k$ where p_k is the pressure difference across the membrane at Q_k, which depends on the internal air pressure and the water depth at Q_k. Thus, given the device geometry, the membrane arc length, and air pressure, we take initial estimates for (T, α) or (T, c) and compute the points Q_k in sequence up to Q_N at (x_N, y_N), say, which is meant to be at the given top attachment point (x_F, y_F), say. By incrementing T and α (or c resp.) by small amounts in turn and computing Q_N again we may deduce an approximation to the Jacobian

$$J = \frac{\partial(x_N, y_N)}{\partial(T, \alpha)} \quad \left(\text{or } \frac{\partial(x_N, y_N)}{\partial(T, c)} \text{ resp.}\right)$$

and set $$\begin{bmatrix} T \\ \alpha \text{ or } c \end{bmatrix}_{j+1} = \begin{bmatrix} T \\ \alpha \text{ or} c \end{bmatrix}_j + J^{-1} \begin{bmatrix} x_F - x_N \\ y_F - y_N \end{bmatrix}$$ iteratively.

For cross-sections near the working position, convergence is often poor unless initial estimates are good. The best technique is to step slowly through a lattice of vectors whose components are the parameters of interest (i.e. air pressure, membrane arc length, water depth, and structure geometry) using the solution in each case to prime the computation for the next neighbouring parameter vector. Fig. 4

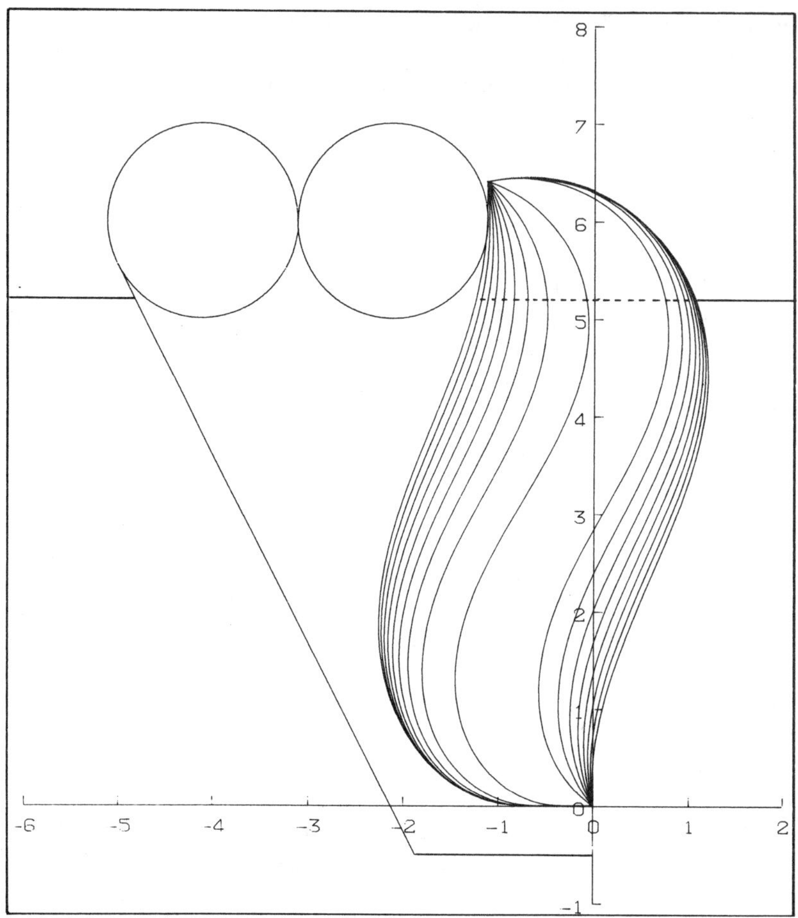

Fig. 4 — Results for a representative CLAM membrane geometry.

shows results for a representative geometry and a range of internal air pressures at equal increments, in water of constant depth. Keeping the air pressure constant and varying the water depth gives a very similar family of shapes. The curves show the low hydrogeometric spring rate (that is the rate of change of air pressure with internal volume) of the membrane system in the mean operation position. (A vertical plate moving horizontally would have a zero spring rate.) A low rate is desirable to match the small added mass component of the hydrodynamic radiation force for the membrane motion, in order to achieve good absorption efficiency of the device at typical wave periods of from 6 to 9 seconds. A earlier version of the CLAM used submerged pillow-shaped air bags mounted on a straight floating spine. Like heaving wedges, these bags have a high spring rate and gave a performance inferior to the membrane type in model tests. Using membranes instead on a straight spine gives stability problems caused by the low spring rate, which were overcome by the change to the torus structure. The geometric information obtained in this way is being used in concurrent work to model the hydrodynamics of the CLAM device based on a full 3-dimensional linear wave diffraction and radiation analysis (Locket 1988).

TROUBLE AT THE MILL

Returning to the general problem of membranes in 3 dimensions, any solution algorithm should ideally be able to cope with arbitrary boundary conditions at edges where the membrane is attached to rigid structure. Progress with the Borrowash device has been severly hampered by development problems with the membrane, notably with regard to its edge attachment, and numerous failures have occurred. Many have been attributed to fitting irregularities which result in sections of membrane with slack cords abutting sections where the cords are taut. The inextensibility of the Kevlar cords used then means that, as loads increase, the taut cords reach their breaking point before they can shed forces to their comparatively unstressed neighbours.

A idealized model of this situation is the unidirectional stressing of a uniform membrane across a finite cut, the objective being to find the augmented stress in the material just beyond the ends of the cut. Consider an infinite plane membrane for which Cartesian axes are principal axes with respect to its uniform elastic properties.

Suppose we tension the membrane uniformly in the y-direction and make a cut along the interval $[-d,d]$ of the x-axis. Let $u(x,y)$ be the displacement of the membrane in the y-direction at (x,y) from its untensioned position and ignore the small stress and strain in the x-direction. The tension force per unit width is then $e\partial u/\partial y$ where e is an elastic modulus and the shear force per unit length is $g\partial u/\partial x$ where g is a shear modulus. From the equilibrium of a small rectangular element we then deduce

$$e\frac{\partial^2 u}{\partial y^2}+g\frac{\partial^2 u}{\partial x^2}=0$$

which gives Laplace's equation on putting $y=Y\sqrt{e}$ and $x=X\sqrt{g}$. The problem is thus the same as that to determine the ideal flow past a plate of width $2d/\sqrt{g}$. Using conformal transformations we may deduce that the complex potential solution Φ satisfies

$$\Phi^2=d^2/g-Z^2,$$

and hence for the membrane problem we have

$$u=\text{Constant.}\quad \mathfrak{R}\sqrt{\left[\frac{d^2}{g}-\left(\frac{x}{\sqrt{g}}+i\frac{y}{\sqrt{e}}\right)\right]}.$$

It follows readily that, along the cut, the displacement is proportional to $\sqrt{\{[(d^2 - x^2)/g]\}}$, so the cut gapes in the form of an ellipse, while across the x-axis beyond the cut we have

$$\text{tension/unit width} = xT_0/\sqrt{[(x^2 - d^2)]}$$

where T_0 is the uniform tension before the cut was made. Integrating over the interval $(d, d+)$ gives

$$\text{tension} = \int_d^{d+\delta} \frac{xT_0 dx}{\sqrt{x^2 - d^2}} = T_0\sqrt{[\delta(2d + \delta)]}$$

When $\delta = 2d/n$ this reduces to $T_0\delta\sqrt{(n + 1)}$ which we may interpret as saying that if the cut extends over n cords, then the first uncut cords experience a tension magnified by a factor $\sqrt{(n + 1)}$.

CONCLUSION

Flexible membranes offer a potentially economic way of harnessing wave energy and provide an interesting challenge to the mathematician to deterine their shape in use. The information obtained will be essential in the process of design and development begun with the Borrowash project, although it now seems unlikely that the hydro device will ever be more cost effective than conventional schemes.

RFEFERENCES

Amirbayat, J. & Hearle, J. W. S. (1985) The complex buckling of flexible sheet materials, Part 1, *Int. J. Mech. Sci.* **28** (6) 339–358.

Lockett, F. P. (1988) Wave forces on the toroidal clam device, *Proc. Euromechanics Colloquium 243 University of Bristol, Sept. 1988.*

12

Continuous, quadilateral, and triangular surface patches

W. S. Hall and T. T. Hibbs
Division of Mathematics and Statistics, School of Information Engineering, Teesside Polytechnic, Middlesbrough, Cleveland, TS13BA

INTRODUCTION

The general represention of surfaces is a subject which has received a large amount of research interest in recent years. Much of this effort has been concerned with C(1) continuous surfaces, particularly in the area of Computer Aided Design (CAD). The definition of C(1) continuity is that the directions of the parametric derivative vectors should be coincident at points on the common boundaries of adjacent elements or patches.

Most of the research on C(1) continuous representations has been carried out in the field of CAD where the objective is to produce a surface with suitable design characteristics. A typical application is the design of a car body, and the research of Bézier (1972) is of fundamental importance in this area. This work entails the creation of a surface rather than the fitting of a surface through specified fixed points, and it uses information at a set of control points. Other applications are concerned with the specification of a C(1) continuous surface which contains a number of fixed nodal points and makes use of information specified at these points. The work of Coons (1964, 1967) and Ferguson (1964) belongs to this category. Unfortunately the surface expressions often partly depend upon nodal information having little physical meaning, such as 'twist' vectors, and are often difficult to implement in practice.

Another important application in engineering is the surface representation of bodies concerned with the solution of problems by the boundary element method. It is desirable that the surface expression should depend only on nodal point values because the same expression is usually used to represent the surface functions. Most implementations of the boundary element method have made use of the shape function representations originally derived by Ergatoudis (1968). These representations use nodal function values only, and they were developed within the context of the finite element method. However, they only provide C(0) continuity which is defined as continuity of the position vector at points on the common boundaries of adjacent elements. Recent research by Hibbs (1989) has shown that it is possible to derive alternative shape function transformations based on nodal function values only which guarantee C(1) continuity at adjacent element boundaries. These transformations were developed specifically for use in the boundary element method, but it is felt that their basic simplicity could make them useful in certain areas of CAD where C(1) continuous surfaces are required.

OVERHAUSER BLENDED CURVES AND THE BASIC SIXTEEN NODE QUADRILATERAL ELEMENT

The derivation of the Overhauser cubic blended curve and its extension to produce a sixteen node quadrilateral element is described by Hall & Hibbs (1988). The results are summarized below for the blended curve and in Fig. 1 for the sixteen node quadrilateral element.

The cubic blended curve is produced by forming a linear blend of two overlapping parabolas which leads to the expression

$$\mathbf{c}(t) = f_1(t)\mathbf{P}_1 + f_2(t)\mathbf{P}_1 + f_3(t)\mathbf{P}_3 + f_4(t)\mathbf{P}_4 \ , \tag{1}$$

$$\begin{aligned} f_1(t) &= -0.5t + t^2 - 0.5t^3 \ , \\ f_2(t) &= 1 - 2.5t^2 + 1.5t^3 \ , \\ f_3(t) &= 0.5t + 2t^2 - 1.5t^3 \ , \\ f_4(t) &= -0.5t^2 + 0.5t^3 \ . \end{aligned} \tag{2}$$

The definition for the blended curve given by equations (1) and (2) makes use of the points $\mathbf{P}_1$ and $\mathbf{P}_4$ outside of the interpolation domain. The functions $f_i(t)$ are the shape functions for the blended curve which is defined between the two inner points $\mathbf{P}_2$ and $\mathbf{P}_3$, with the parameter t contained in the interval [0,1]. The basic construction of blended curves using overlapping parabolas leads to the fundamental property that adjacent blended curves are C(1) continuous at their joining nodes. The important fact is that this C(1) continuity has been achieved without the use of derivative nodal freedoms. The basic sixteen-node quadrilateral element is shown in Fig. 1, and it is constructed from two separate surfaces where both surfaces are transformed to the unit square. The first surface $\mathbf{S}(u,v)$ is constructed by defining the four vertical blended curves $\mathbf{d}_1(v)$ to $\mathbf{d}_4(v)$ and using these curves to calculate, for any v, the four points $\mathbf{S}_1$ to $\mathbf{S}_4$. These four points are used to form the surface

$$\mathbf{S}(u,v) = f_1(u)\mathbf{S}_1 + f_2(u)\mathbf{S}_2 + f_3(u)\mathbf{S}_3 + f_4(u)\mathbf{S}_4 \ . \tag{3}$$

The fundamental property of Overhauser blended curves guarantees that the surface $\mathbf{S}(u,v)$ will be C(1) continuous with adjacent surfaces along the parameter lines $u = 0$ and $u = 1$.

The second surface $\mathbf{T}(u,v)$ is constructed by defining the four horizontal blended curves $\mathbf{c}_1(u)$ to $\mathbf{c}_4(u)$ and calculating, for any u, the four points $\mathbf{T}_1$ to $\mathbf{T}_4$. These four points are used to form the surface

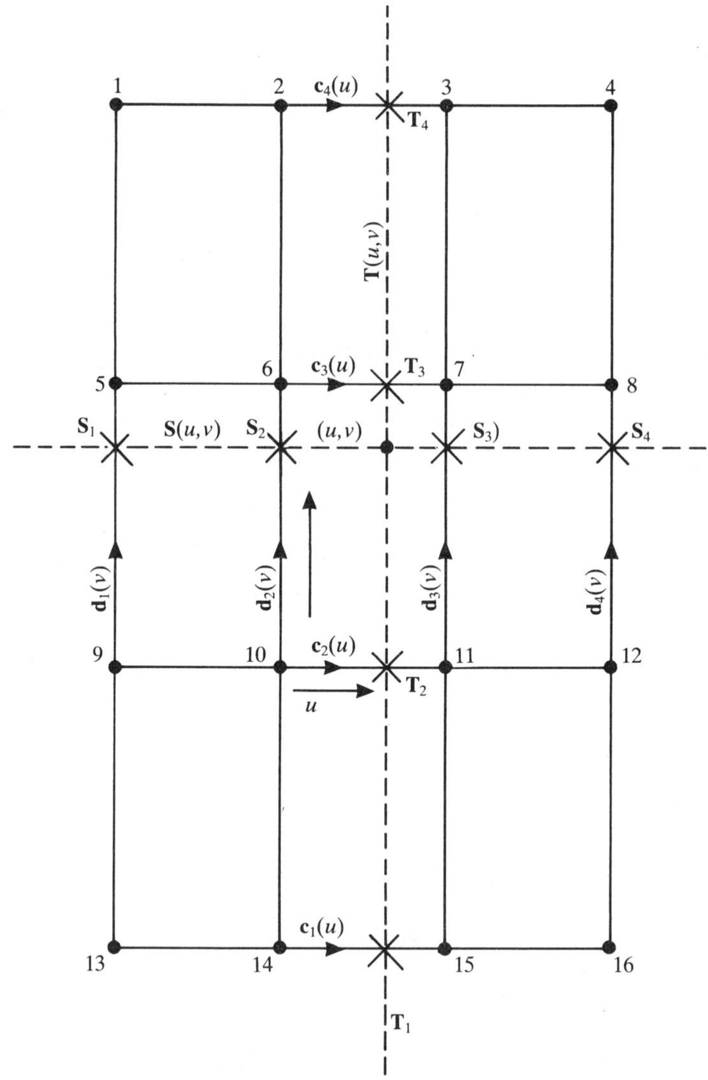

Fig. 1 — The sixteen node Overhauser quadrilateral element.

$$\mathbf{T}(u,v) = f_1(v)\mathbf{T}_1 + f_2(v)\mathbf{T}_2 + f_3(v)\mathbf{T}_3 + f_4(v)\mathbf{T}_4 \ . \tag{4}$$

Again the fundamental property of Overhauser blended curves ensures that the surface $\mathbf{T}(u,v)$ will be C(1) continuous with adjacent surfaces along the parameter lines $v = 0$ and $v = 1$.

By substituting the blended curve definitions for the points $\mathbf{S}_1$ to $\mathbf{S}_4$ and $\mathbf{T}_1$ to $\mathbf{T}_4$ in equations (3) and (4) it is found that the expressions for the surfaces $\mathbf{S}(u,v)$ and $\mathbf{T}(u,v)$ are identical and may be written in the form

$$\mathbf{Q}(u,v) = \sum_{i=1}^{16} w_i(u,v)\mathbf{P}_i \tag{5}$$

where the nodal weight functions $w_i(u,v)$ are defined as products of the blended cure shape functions $f_i(t)$. Thus a surface has been found which is C(1) continuous at its edges will all other surfaces defined in the same way. It should be noted that the nodal weights $w_i(u,v)$ in equation (5) satisfy the shape function definition.

A C(1) CONTINUOUS TRIANGULAR ELEMENT

The C(1) continuous quadrilateral element described in the previous section is not sufficient to allow a suitable mesh to be designed for all surfaces. This is because it requires that all elements must meet four to a corner, which is essential in the construction of the quadrilateral element. There is thus a need for a triangular element which is C(1) continuous with the basic quadrilateral element which is C(1) continuous with the basic quadrilateral element developed previously. To ensure this C(1) continuity requires the use of special elements, called transition elements, which are described in the next section.

The form of the triangular element is shown in Fig. 2, and it requires twelve nodes for its definition. The principles, which were developed during the derivation of the quadrilateral element, are applied to the triangular element in this section. The main difference between the two applications is that three component surfaces are required for the triangular element. The construction of the first surface $\mathbf{Q}_1(L_1,t_1)$ is shown in Fig. 2 where the parameters L_1 and t_1 are functions of the general parametric position (u,v). The parameter L_1 is the area coordinate and the parameter t_1 is the side parameter for the component surface $\mathbf{Q}_1(L_1,t_1)$ which is constructed from the four points $\mathbf{R}_1$, $\mathbf{R}_2$, $\mathbf{P}_3$, and $\mathbf{R}_4$. The points $\mathbf{R}_1$ and $\mathbf{R}_2$ are found from the blended curves $\mathbf{c}_1(t_1)$ and $\mathbf{c}_2(t_1)$ which are defined by the nodal positions $\mathbf{P}_5$ to $\mathbf{P}_{12}$. The point $\mathbf{R}_4$ is found from the parabola defined by the nodal positions $\mathbf{P}_2$, P_1, and $\mathbf{P}_4$ as

$$\mathbf{d}_1(t_1) = h_1(t_1)\mathbf{P}_2 + h_2(t_1)\mathbf{P}_1 + h_3(t_1)\mathbf{P}_4 \ . \tag{6}$$

$$h_1(t) = 1 - 3t + 2t^2 \ ,$$

$$h_2(t) = 4t - 4t^2 \ ,$$

$$h_3(t) = -t + 2t^2 \ . \tag{7}$$

The functions $h_i(t)$ are the shape functions for the parabolic curve.

The components surface $\mathbf{Q}_1(L_1,t_1)$ is constructed by forming a blended curve from the four positions $\mathbf{R}_1$, $\mathbf{R}_2$, $\mathbf{P}_3$, and $\mathbf{R}_4$ as

$$\mathbf{Q}_1(L_1,t_1) = f_1(L_1)\mathbf{R}_1 + f_2(L_1)\mathbf{R}_2 + f_3(L_1)\mathbf{P}_3 + f_4(L_1)\mathbf{R}_4 \tag{8}$$

where the position vectors $\mathbf{R}_1$, $\mathbf{R}_2$, and $\mathbf{R}_4$ are functions of the side parameter t_1. By appealing to the symmetry in the construction of the triangular element the two remaining component surfaces may be written as

$$\mathbf{Q}_2(L_2,t_2) = f_1(L_2)\mathbf{S}_1 + f_2(L_2)\mathbf{S}_2 + f_3(L_2)\mathbf{P}_6 + f_4(L_2)\mathbf{S}_4 \ , \tag{9}$$

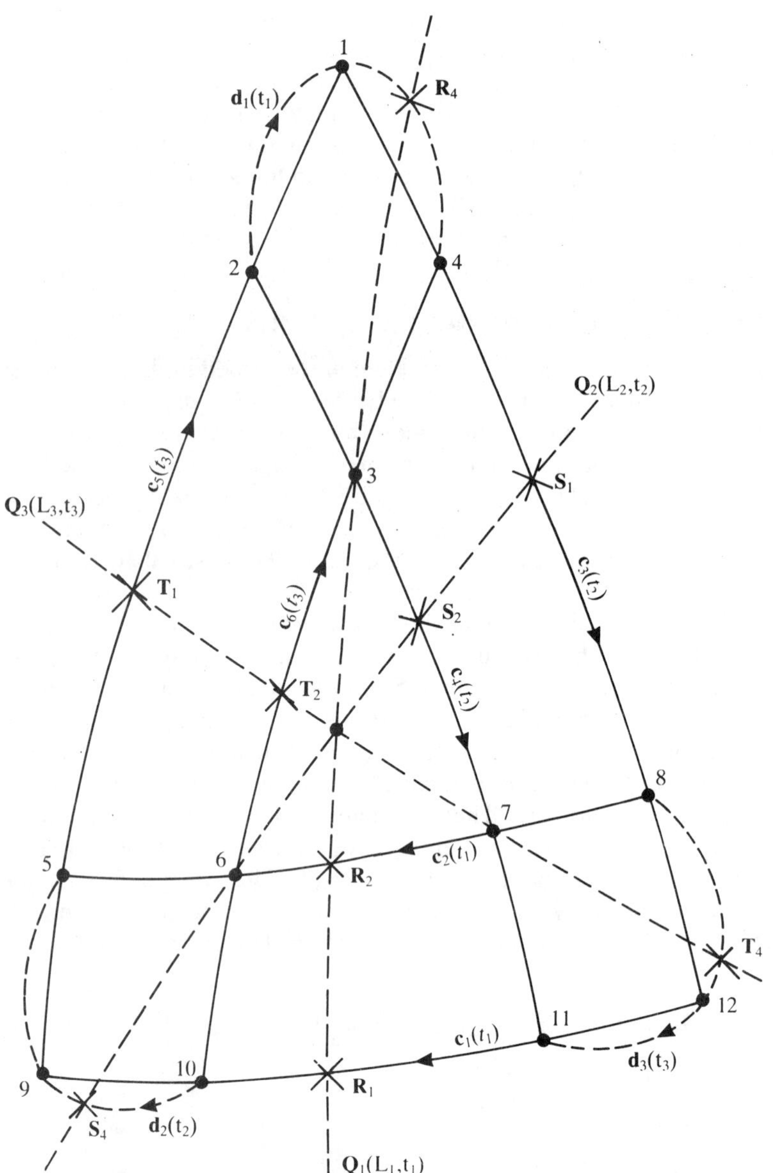

Fig. 2 — The C(1) continuous triangular element.

$$\mathbf{Q}_3(L_2,t_3) = f_1(L_3)\mathbf{T}_1 + f_2(L_3)\mathbf{T}_2 + f_3(L_3)\mathbf{P}_7 + f_4(L_3)\mathbf{T}_4 \ , \tag{10}$$

where the position vectors $\mathbf{S}_1$, $\mathbf{S}_2$, and $\mathbf{S}_4$ are functions of the side parameter t_2 and the position vectors $\mathbf{T}_1$, $\mathbf{T}_2$ and $\mathbf{T}_4$ are functions of the side parameter t_3.

Substitution of the definitions for the blended curves $\mathbf{c}_i(t)$ and the vertex parabolas $\mathbf{d}_i(t)$ into equations (8) to (10) gives the final expressions for the component surfaces $\mathbf{Q}_i(L_i,t_i)$ which are all different in this case. The final surface expression is found by forming a convex combination of the component surfaces in such a way that C(1) continuity is obtained at the common boundaries with adjacent quadrilateral elements as suggested by Gregory & Charrott (1980). The final surface expression for the triangular element is thus given by

$$\mathbf{Q}(L_i,t_i) = A_1(L_i)\mathbf{Q}_1(L_1,t_1) + A_2(L_i)\mathbf{Q}_2(L_2,t_2) + A_3(L_i)\mathbf{Q}_3(L_3,t_3) \tag{11}$$

where the blending coefficients $A_1(L_i)$, $A_2(L_i)$, and $A_3(L_i)$ have the following convexity properties:

$$A_1(L_i) \geq 0 \ , \qquad A_2(L_i) \geq 0 \ , \qquad A_3(L_i) \geq 0 \ ,$$
$$A_1(L_i) + A_2(L_i) + A_3(L_i) = 1 \ . \tag{12}$$

Additional properties are required for the blending coefficients in this case to ensure that the correct component surface is used at points on the boundary thus ensuring C(1) continuity with adjacent quadrilateral elements. The additional properties are:

$$A_1(L_i) = 1 \text{ at } L_1 = 0 \ , \qquad A_2(L_i) = 1 \text{ at } L_2 = 0 \ ,$$
$$A_3(L_i) = 1 \text{ at } L_3 = 0 \ . \tag{13}$$

The derivative properties required are that all of the first derivatives of the blending coefficients are zero on the boundary of the triangular element. A complete discussion of the derivation of the triangular element, including the choice of the blending coefficients, is given by Hibbs (1989).

The final expression for the triangular element is found by substituting equations (8) to (10) into equation (11) and by replacing the points $\mathbf{R}_1$, $\mathbf{R}_2$, $\mathbf{R}_4$, $\mathbf{S}_1$, $\mathbf{S}_2$, $\mathbf{S}_4$, $\mathbf{T}_1$, $\mathbf{T}_2$, and $\mathbf{T}_4$ by the expressions for their defining curves. This expression can be written in the form:

$$\mathbf{Q}(u,v) = \sum_{i=1}^{12} w_i(u,v)\mathbf{P}_i \tag{14}$$

where the nodal weight functions $w_i(u,v)$ satisfy the shape function definition.

QUADRILATERAL TRANSITION ELEMENTS

Triangular elements, as developed in the previous section, must be associated with special quadrilateral elements to blend them into an otherwise quadrilateral mesh. Fig. 3 shows a triangular element marked A surrounded by a quadrilateral mesh, and it can be seen that only two types of transitions element are necessary. These elements are a common side transition element marked B and a common node transition element marked C. The quadrilateral transition elements are constructed in a similar manner to the basic sixteen-node quadrilateral element, but in this case the two component surfaces are not identical. This means that the component surfaces $\mathbf{S}(u,v)$ and $\mathbf{T}(u,v)$ must be combined by using suitable blending coefficients $A(u,v)$ and $B(u,v)$ to form the final surface expression given by

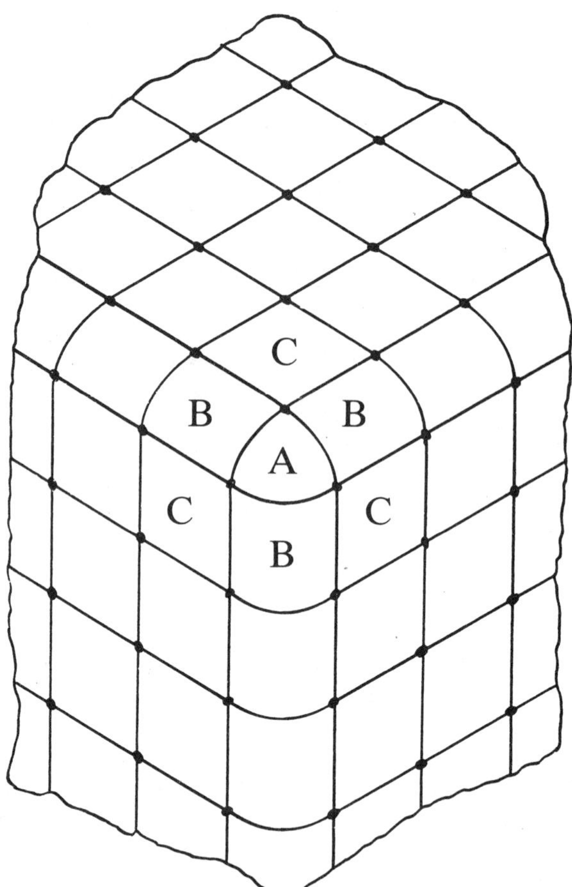

Fig. 3 — Types of transition element.

$$\mathbf{Q}(u,v) = A(u,v)\mathbf{S}(u,v) + B(u,v)\mathbf{T}(u,v) \tag{15}$$

where the properties of the blending coefficient are given by

$$A(u,v) \geqslant 0 \;, \qquad B(u,v) \geqslant 0 \;, \qquad A(u,v) + B(u,v) = 1 \;,$$
$$A(u,v) = 1 \text{ when } u = 0 \text{ and } u = 1 \;,$$
$$B(u,v) = 1 \text{ when } v = 0 \text{ and } v = 1 \;. \tag{16}$$

In addition, all the first derivatives of the blending coefficients are zero on the boundary of the quadrilateral element.

Full details of the derivation of both types of quadrilateral transition element can be found in Hibbs (1989) where it is shown that the final expression for both types of transition element may be written in shape function form.

REDUCED QUADRILATERAL ELEMENTS

For practical applications, where surfaces have edges and corners, it is necessary to define reduced versions of the quadrilateral elements. These reduced elements are required because it is not possible to fully specify the list of nodes which define the element when it is situated near to the edge or a corner. The reduced version of a particular quadrilateral element is obtained by using quadratic extrapolation to predict the position of the missing nodes and then applying the basic element expression. This approach is used because the reduced element is still defined in terms of nodal point values situated on the surface. This property is very desirable for applications of the element in the Boundary Element Method where the same expressions are usually used to model the geometry and the surface functions. For applications in Computer Aided Design it may be possible to specify the full list of nodes defining a quadrilateral element by making use of function values specified at nodes not situated on the surface. This approach would mean that reduced elements are not required. A full description of all the reduced elements needed is given in Hibbs (1989) where the reduced element expressions are given in shape function form.

NUMERICAL RESULTS

Some numerical results are presented in Table 1 for two adjacent sixteen-node quadrilateral test elements A and B situated on the surface of a sphere of radius twenty units. The sixteen nodal position vectors necessary to define each element on the spherical surface are given by the intersection of four lines of longitude U_i with four lines of latitude V_j. The test element A is defined by the parameters values

$$U_i = 10,\ 20,\ 30,\ \text{and } 40 \text{ degrees},$$

$$V_j = 10,\ 20,\ 30,\ \text{and } 40 \text{ degrees}$$

and the test element B is defined by the parameter values

$$U_i = 20,\ 30,\ 40,\ \text{and } 50 \text{ degrees},$$

$$V_j = 10,\ 20,\ 30,\ \text{and } 40 \text{ degrees}.$$

By using the exact transformation for a spherical surface it is possible to make comparisons between exact and approximate results for position and derivative vectors at chosen points within the test elements. In Table 1 the entries labelled 'approximate' refer to results obtained using the representation defined by equation (5), and the entries labelled 'exact' refer to exact values on the sphere. The results are given in the order of position vector $\mathbf{x}$, derivative vector in the u parameter direction $\mathbf{x}_u$, the derivative vector in the v parameter direction $\mathbf{x}_v$ respectively.

Results are given for two sample points **a** and **b** with parameter coordinates

$$\mathbf{a};\quad (0.5\ ,\ 0.5)\ ,\qquad \mathbf{b};\quad (1.0\ ,\ 0.5)$$

in test element A and for two sample points **b*** and **c** with parameter coordinates

$$\mathbf{b}^*;\quad (0.0\ ,\ 0.5)\ ,\qquad \mathbf{c};\quad (0.5\ ,\ 0.5)$$

Table 1 — Results for test elements A and B on the sphere

		Approximate values			Exact values		
a	$\mathbf{x}$	7.6601	3.5720	18.1258	7.6604	3.5721	18.1262
	$\mathbf{x}_u$	2.8744	1.3404	−1.4789	2.8672	1.3370	−1.4752
	$\mathbf{x}_v$	−0.6250	1.3404	0.0	−0.6235	1.3370	0.0
b	$\mathbf{x}$	9.0629	4.2261	17.3205	9.0631	4.2262	17.3205
	$\mathbf{x}_u$	2.7258	1.2711	−1.7365	2.7398	1.2776	−1.7453
	$\mathbf{x}_v$	−0.7395	1.5858	0.0	−0.7376	1.5818	0.0
b*	$\mathbf{x}$	9.0629	4.2261	17.3205	9.0631	4.2262	17.3205
	$\mathbf{x}_u$	2.7258	1.2711	−1.7365	2.7398	1.2776	−1.7453
	$\mathbf{x}_v$	−0.7395	1.5858	0.0	−0.7376	1.5818	0.0
c	$\mathbf{x}$	10.3963	4.8479	16.3827	10.3967	4.8481	16.3830
	$\mathbf{x}_u$	2.5980	1.2115	−2.0072	2.5915	1.2084	−2.0022
	$\mathbf{x}_v$	−0.8483	1.8191	0.0	−0.8462	1.8146	0.0

in test element B. Of course, the sample points **b** and **b*** are coincident points for test elements A and B.

Table 1 shows that the approximate results and the exact results generally agree to three significant figures. This represents a good measure of agreement and demonstates the accuracy that can be obtained in modelling curved surfaces using nodal position vectors only.

REFERENCES

Bézier, P. (1972) *Numerical Control, Mathematics and Applications*, Wiley.

Coons, S. A. (1964) *Surfaces for Computer Aided Design of Space Figures*, M.I.T. Report ESL 9442-M-139.

Coons, S. A. (1967) *Surfaces for Computer Aided Design of Space Forms*, M.I.T. Report MAC-TR-41.

Ergatoudis, J. G. (1968) Isoparametric finite elements in two and three dimensional stress analysis, PhD thesis, University of Wales.

Ferguson, J. C. (1964) Multivariable curve interpolation, *J. Assoc. Comp. Mach.* **11** 221–228.

Gregory, J. A. & Charrott, P. (1980) A C(1) continuous triangular interpolation patch for computer aided geometric design, *Comp. Graphics Image Process.* **13** 80–87.

Hall, W. S. & Hibbs, T. T. (1988) C(1) continuous, quadrilateral boundary elements applied to three dimensional problems in potential theory. In: *Boundary Elements X*, **2**, ed. C. A. Brebbia, Springer-Verlag.

Hibbs, T. T. (1989) C(1) continuous representations and advanced singular kernel integrations in the three dimensional boundary integral method, PhD thesis, CNAA, London.

13

Overhauser patches for generally curved surfaces

D. J. M. Hill, E. J. Fletcher, and **A. O. Moscardini**
School of Computer Studies and Mathematics, Sunderland Polytechnic
T. S. Wilkinson
Mathematical Services Dept, NEI Parsons Ltd, Newcastle upon Tyne

INTRODUCTION

As part of the development of a software package aimed at aiding the design, analysis, and manufacture of turbine blades, the writers have considered the use of a number of generally curved surface patches which together completely define the surface of a solid object.

Surface elements based on Overhauser (see Overhauser 1968) blended curves have been used recently at Teeside Polytechnic (Hall & Hibbs 1987) in, for example, a 3-D elastic stress analysis program using the boundary element method. They require a minimum of data (other than surface coordinates) and their surface gradients are continuous at the junction between adjacent patches. For these reasons, they were selected for further study.

In this chapter, Overhauser curve segments and surface patches will be defined from first principles. A vector/matrix equation will then be derived for the surface patch to enable comparison with other, well-known bicubic patches. It will then be deduced that the Overhauser surface patch is, in fact, a Ferguson patch (Faux & Pratt 1979), with centrally estimated first partial derivatives and twist vectors set to be Adini's twists (Barnhill *et al.* 1976). Other researchers, having examined various twist estimation methods for Ferguson patches, have concluded that Adini's twists give the best result.

FORMULATION OF OVERHAUSER BLENDED CURVES AND PATCHES

An Overhauser curve is a cubic polynomial segment formed by taking a linear blend of two overlapping parabolas. Thus, considering the four points $\mathbf{p}_1$, $\mathbf{p}_2$, $\mathbf{p}_3$, $\mathbf{p}_4$ in Fig. 1, the blended curve $\mathbf{r}(t)$ will be defined between the central points $\mathbf{p}_2$ and $\mathbf{p}_3$ as:

$$\mathbf{r}(t) = (1-t)\mathbf{p}(\alpha) + t\mathbf{q}(\beta) \qquad 0 \leqslant t \leqslant 1 \tag{1}$$

where $\mathbf{p}$ and $\mathbf{q}$ are quadratic polynomial segments through $\mathbf{p}_1$, $\mathbf{p}_2$, $\mathbf{p}_3$, and $\mathbf{p}_2$, $\mathbf{p}_3$, $\mathbf{p}_4$ respectively and α and β are linear functions of t. If the parabolas $\mathbf{p}$ and $\mathbf{q}$ are chosen such that $\mathbf{p}_2$ and $\mathbf{p}_3$ are the respective parametric midpoints (that is, $\mathbf{p}(\frac{1}{2}) = \mathbf{p}_2$, $\mathbf{q}(\frac{1}{2}) = \mathbf{p}_3$) then

$$\mathbf{p}(\alpha) = [\alpha^2 \quad \alpha \quad 1]\begin{bmatrix} 2 & -4 & 2 \\ -3 & 4 & -1 \\ 1 & 0 & 0 \end{bmatrix}\begin{bmatrix} \mathbf{p}_1 \\ \mathbf{p}_2 \\ \mathbf{p}_3 \end{bmatrix} \qquad 0 \leqslant \alpha \leqslant 1$$

$$\mathbf{q}(\beta) = [\beta^2 \quad \beta \quad 1]\begin{bmatrix} 2 & -4 & 2 \\ -3 & 4 & -1 \\ 1 & 0 & 0 \end{bmatrix}\begin{bmatrix} \mathbf{p}_2 \\ \mathbf{p}_3 \\ \mathbf{p}_4 \end{bmatrix} \qquad 0 \leqslant \beta \leqslant 1 \tag{2}$$

Since $\mathbf{p}(\frac{1}{2}) = \mathbf{r}(0) = \mathbf{q}(0)$ and $\mathbf{p}(1) = \mathbf{r}(1) = \mathbf{q}(\frac{1}{2})$ then

$$\alpha = \tfrac{1}{2}(t+1), \quad \beta = \tfrac{1}{2}t.$$

Substituting these expressions for α and β into (2) and equations (2) into (1),

$$\mathbf{r}(t) = (1-t)[\tfrac{1}{4}(t+1)^2 \quad \tfrac{1}{2}(t+1) \quad 1]\begin{bmatrix} 2 & -4 & 2 \\ -3 & 4 & -1 \\ 1 & 0 & 0 \end{bmatrix}\begin{bmatrix} \mathbf{p}_1 \\ \mathbf{p}_2 \\ \mathbf{p}_3 \end{bmatrix}$$

$$+ t[\tfrac{1}{4}t^2 \quad \tfrac{1}{2}t \quad 1]\begin{bmatrix} 2 & -4 & 2 \\ -3 & 4 & -1 \\ 1 & 0 & 0 \end{bmatrix}\begin{bmatrix} \mathbf{p}_2 \\ \mathbf{p}_3 \\ \mathbf{p}_4 \end{bmatrix}$$

$$= [t^3 \quad t^2 \quad t \quad 1]\begin{bmatrix} -\frac{1}{2} & \frac{3}{2} & -\frac{3}{2} & 2 \\ 1 & -\frac{5}{2} & -2 & -\frac{1}{2} \\ -\frac{1}{2} & 0 & \frac{1}{2} & 0 \\ 0 & 1 & 0 & 0 \end{bmatrix}\begin{bmatrix} \mathbf{p}_1 \\ \mathbf{p}_2 \\ \mathbf{p}_3 \\ \mathbf{p}_4 \end{bmatrix} \tag{3}$$

Immediately, these ideas can be extended to form composite curves, using successive overlapping parabolas (Fig. 2).

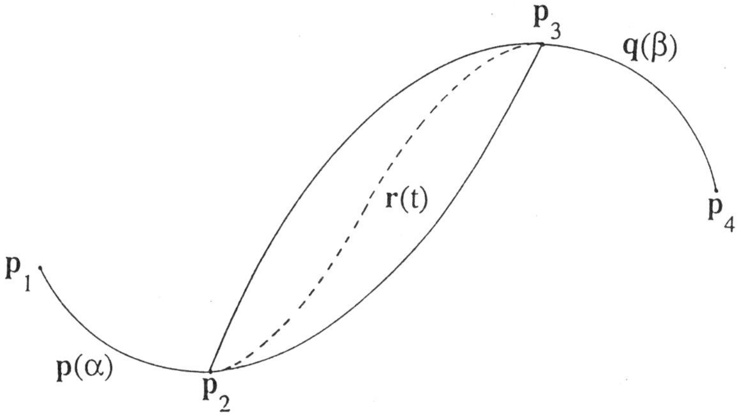

Fig. 1 — Overhauser curved segment.

Using the fact that for the curve $\mathbf{r}(t)$ above

$$\dot{\mathbf{r}}(0) = \tfrac{1}{2}(\mathbf{p}_3 - \mathbf{p}_1)$$
$$\dot{\mathbf{r}}(1) = \tfrac{1}{2}(\mathbf{p}_4 - \mathbf{p}_2),$$

it can be deduced that the composite curve is in fact C(1) continuous.

Given a surface specified by a quadrilateral arrangement of points, the above ideas can be exrended to form an Overhauser surface patch. Consider the patch in Fig. 3 whose corners are $\mathbf{a}_{22}$, $\mathbf{a}_{23}$, $\mathbf{a}_{32}$, and $\mathbf{a}_{33}$.

To define the point $\mathbf{r}(u, v)$ $0 \leq u \leq 1$, $0 \leq v \leq 1$, the points $\mathbf{p}_1$, $\mathbf{p}_2$, $\mathbf{p}_3$, and $\mathbf{p}_4$ are generated by defining Overhauser curves using the points $\mathbf{a}_{k1}$, $\mathbf{a}_{k2}$, $\mathbf{a}_{k3}$, and $\mathbf{a}_{k4}$ for the point $\mathbf{p}_k$. That is,

$$\mathbf{p}_k = [u^3 \quad u^2 \quad u \quad 1] \begin{bmatrix} -\frac{1}{2} & \frac{3}{2} & -\frac{3}{2} & \frac{1}{2} \\ 1 & -\frac{5}{2} & 2 & -\frac{1}{2} \\ -\frac{1}{2} & 0 & \frac{1}{2} & 0 \\ 0 & 1 & 0 & 0 \end{bmatrix} \begin{bmatrix} \mathbf{a}_{k1} \\ \mathbf{a}_{k2} \\ \mathbf{a}_{k3} \\ \mathbf{a}_{k4} \end{bmatrix} \quad k = 1, 2, 3, 4 \tag{4}$$

or, equivalently,

$$\mathbf{p}_k = \sum_{i=1}^{4} \sum_{j=1}^{4} u^{4-i} \mathbf{q}_{ij} \mathbf{a}_{kj}$$

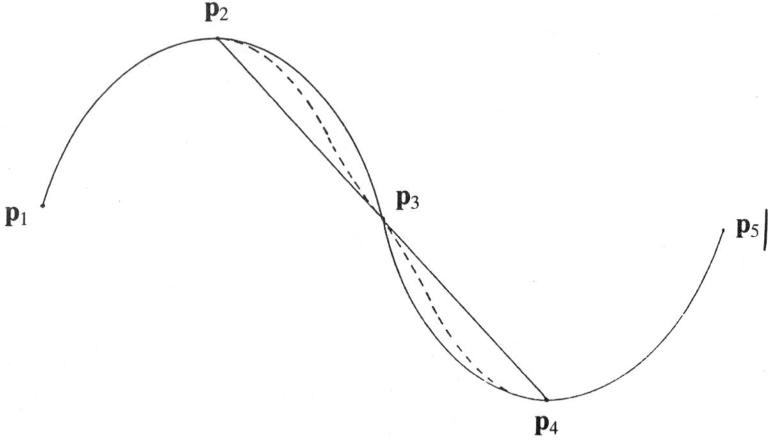

Fig. 2 — Composite Overhauser curve.

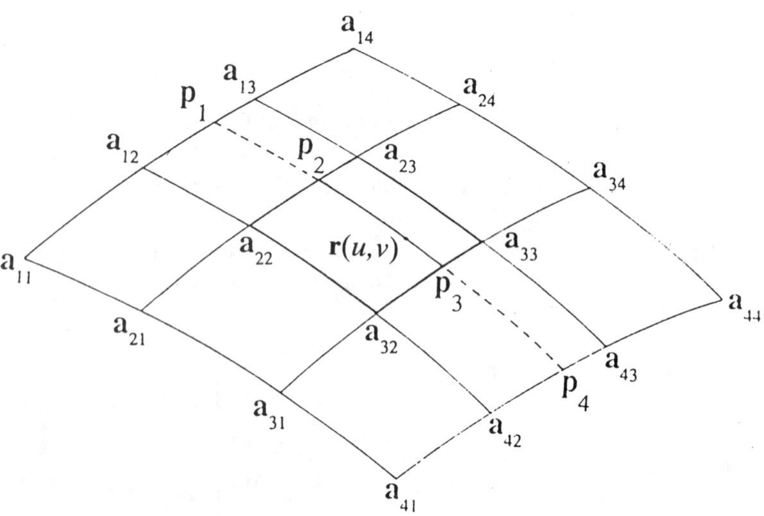

Fig. 3 — Overhauser surface patch.

where the matrix $Q = [\mathbf{q}_{ij}]_{i,j=1}^{4} = \begin{bmatrix} -\frac{1}{2} & \frac{3}{2} & -\frac{3}{2} & \frac{1}{2} \\ 1 & -\frac{5}{2} & 2 & -\frac{1}{2} \\ -\frac{1}{2} & 0 & \frac{1}{2} & 0 \\ 0 & 1 & 0 & 0 \end{bmatrix}$.

The points $\mathbf{p}_1$, $\mathbf{p}_2$, $\mathbf{p}_3$, and $\mathbf{p}_4$ are then used to form a fifth Overhauser curve, which gives $\mathbf{r}(u, v)$:

$$\mathbf{r}(u,v) = [v^3 \quad v^2 \quad v \quad 1] \begin{bmatrix} -\frac{1}{2} & \frac{3}{2} & -\frac{3}{2} & \frac{1}{2} \\ 1 & -\frac{5}{2} & 2 & -\frac{1}{2} \\ -\frac{1}{2} & 0 & \frac{1}{2} & 0 \\ 0 & 1 & 0 & 0 \end{bmatrix} \begin{bmatrix} \mathbf{p}_1 \\ \mathbf{p}_2 \\ \mathbf{p}_3 \\ \mathbf{p}_4 \end{bmatrix} \tag{5}$$

The extension to form composite surfaces is done in an analogous manner to composite curves. Again, it can be shown that the surface formed is C(1) continuous. Also, the above definition of $\mathbf{r}(u, v)$ is well-defined in the sense that using the points $\mathbf{a}_{1k}$, $\mathbf{a}_{2k}$, $\mathbf{a}_{3k}$, and $\mathbf{a}_{4k}$ to generate the $\mathbf{p}_k$ leads to the same point $\mathbf{r}(u, v)$.

COMPARISON WITH THE FERGUSON PATCH

According the Faux and Pratt (1979), the Ferguson patch (Fig. 4) is given by the

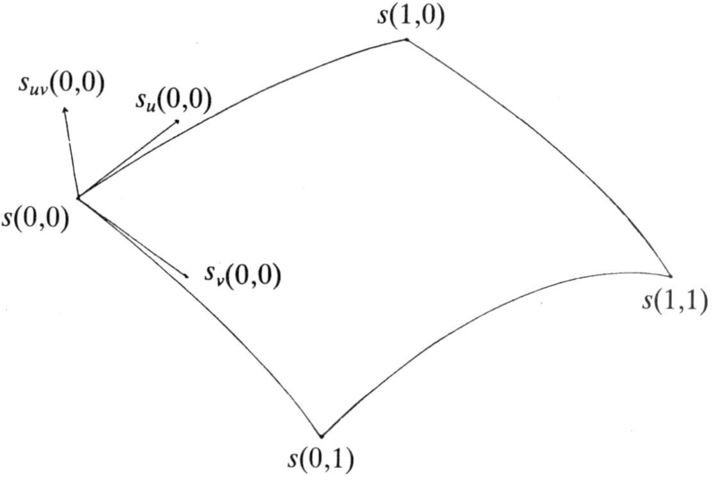

Fig. 4 — Ferguson surface patch.

following vector/matrix equation:

$$\mathbf{r}(u,v) = [v^3 \quad v^2 \quad v \quad 1]\mathrm{FSF^T}[u^3 \quad u^2 \quad u \quad 1]^{\mathrm{T}} \tag{6}$$

where

$$F = \begin{bmatrix} 2 & -2 & 1 & 1 \\ -3 & 3 & -2 & -1 \\ 0 & 0 & 1 & 0 \\ 1 & 0 & 0 & 0 \end{bmatrix}$$

and

$$S = \begin{bmatrix} \mathbf{s}(0,0) & \mathbf{s}(1,0) & \mathbf{s}_u(0,0) & \mathbf{s}_u(1,0) \\ \mathbf{s}(0,1) & \mathbf{s}(1,1) & \mathbf{s}_u(0,1) & \mathbf{s}_u(1,1) \\ \mathbf{s}_\upsilon(0,0) & \mathbf{s}_\upsilon(1,0) & \mathbf{s}_{u\upsilon}(0,0) & \mathbf{s}_{u\upsilon}(1,0) \\ \mathbf{s}_\upsilon(0,1) & \mathbf{s}_\upsilon(1,1) & \mathbf{s}_{u\upsilon}(0,1) & \mathbf{s}_{u\upsilon}(1,1) \end{bmatrix}$$

To compare the Ferguson and the Overhauser patches, a similar vector/matrix equation must first be formulated.

Equation (5) may be rewritten as

$$\mathbf{r}(u,\upsilon) = \sum_{k=1}^{4}\sum_{l=1}^{4} \upsilon^{4-k}\mathbf{q}_{kl}\mathbf{p}_1$$

Substituting from equation (4) for $\mathbf{p}_1$,

$$= \sum_{k=1}^{4}\sum_{l=1}^{4} \upsilon^{4-k}\mathbf{q}_{kl}\sum_{i=1}^{4}\sum_{j=1}^{4} u^{4-i}\mathbf{q}_{ij}\,\mathbf{a}_{lj}$$

$$= \sum_{i=1}^{4}\sum_{j=1}^{4}\sum_{k=1}^{4}\sum_{l=1}^{4} \upsilon^{4-k}\mathbf{q}_{ij}\mathbf{a}_{lj}q_{kl}u^{4-i}$$

$$= \sum_{k=1}^{4}\sum_{i=1}^{4} \upsilon^{4-k}\left[\sum_{l=1}^{4}\sum_{j=1}^{4} \mathbf{q}_{kl}\mathbf{a}_{lj}\mathbf{q}_{ij}\right]u^{4-i}$$

$$= [\upsilon^3 \quad \upsilon^2 \quad \upsilon \quad 1]\mathrm{QAQ}^{\mathrm{T}}[u^3 \quad u^2 \quad u \quad 1]^{\mathrm{T}} \tag{7}$$

where the matrix $\mathrm{A} = [\mathbf{a}_{ij}]_{i,j=1}^{4}$.

Now,

$$Q = \begin{bmatrix} -\frac{1}{2} & \frac{3}{2} & -\frac{3}{2} & \frac{1}{2} \\ 1 & -\frac{5}{2} & 2 & -\frac{1}{2} \\ -\frac{1}{2} & 0 & \frac{1}{2} & 0 \\ 0 & 1 & 0 & 0 \end{bmatrix} = \begin{bmatrix} 2 & -2 & 1 & 1 \\ -3 & 3 & -2 & -1 \\ 0 & 0 & 1 & 0 \\ 1 & 0 & 0 & 0 \end{bmatrix} \begin{bmatrix} 0 & 1 & 0 & 0 \\ 0 & 0 & 1 & 0 \\ -\frac{1}{2} & 0 & \frac{1}{2} & 0 \\ 0 & -\frac{1}{2} & 0 & \frac{1}{2} \end{bmatrix}$$

$$= \mathrm{FG}, \text{ say.}$$

Hence equation (7) becomes:

$$\begin{aligned} \mathbf{r}(u,v) &= [v^3 \quad v^2 \quad v \quad 1]\mathrm{FGAG^TF^T}[u^3 \quad u^2 \quad u \quad 1]^\mathrm{T} \\ &= [v^3 \quad v^2 \quad v \quad 1]\mathrm{FHF^T}[u^3 \quad u^2 \quad u \quad 1]^\mathrm{T} \end{aligned}$$

where the matrix

$$\mathrm{H} = \begin{bmatrix} \mathbf{a}_{22} & \mathbf{a}_{23} & \frac{1}{2}(\mathbf{a}_{23}-\mathbf{a}_{21}) & \frac{1}{2}(\mathbf{a}_{24}-\mathbf{a}_{22}) \\ \mathbf{a}_{32} & \mathbf{a}_{33} & \frac{1}{2}(\mathbf{a}_{33}-\mathbf{a}_{31}) & \frac{1}{2}(\mathbf{a}_{34}-\mathbf{a}_{32}) \\ \frac{1}{2}(\mathbf{a}_{32}-\mathbf{a}_{12}) & \frac{1}{2}(\mathbf{a}_{33}-\mathbf{a}_{13}) & \begin{matrix}\frac{1}{4}(\mathbf{a}_{33}-\mathbf{a}_{13}) \\ -\frac{1}{4}(\mathbf{a}_{31}-\mathbf{a}_{11})\end{matrix} & \begin{matrix}\frac{1}{4}(\mathbf{a}_{34}-\mathbf{a}_{14}) \\ -\frac{1}{4}(\mathbf{a}_{32}-\mathbf{a}_{12})\end{matrix} \\ \frac{1}{2}(\mathbf{a}_{42}-\mathbf{a}_{22}) & \frac{1}{2}(\mathbf{a}_{43}-\mathbf{a}_{23}) & \begin{matrix}\frac{1}{4}(\mathbf{a}_{43}-\mathbf{a}_{23}) \\ -\frac{1}{4}(\mathbf{a}_{41}-\mathbf{a}_{21})\end{matrix} & \begin{matrix}\frac{1}{4}(\mathbf{a}_{44}-\mathbf{a}_{24}) \\ -\frac{1}{4}(\mathbf{a}_{42}-\mathbf{a}_{22})\end{matrix} \end{bmatrix}.$$

Comparing this equation with that for the Ferguson patch (6), it can be deduced that the Overhauser patch is in fact a Ferguson patch with centrally estimated first partial derivatives and the cross-derivatives, or 'twists', set to be:

$$\begin{aligned} s_{uv}(0,0) &= \tfrac{1}{4}(\mathbf{a}_{33}-\mathbf{a}_{13}) - \tfrac{1}{4}(\mathbf{a}_{31}-\mathbf{a}_{11}) \\ s_{uv}(0,1) &= \tfrac{1}{4}(\mathbf{a}_{43}-\mathbf{a}_{23}) - \tfrac{1}{4}(\mathbf{a}_{41}-\mathbf{a}_{21}) \\ s_{uv}(1,0) &= \tfrac{1}{4}(\mathbf{a}_{34}-\mathbf{a}_{14}) - \tfrac{1}{4}(\mathbf{a}_{32}-\mathbf{a}_{12}) \\ s_{uv}(1,1) &= \tfrac{1}{4}(\mathbf{a}_{44}-\mathbf{a}_{24}) - \tfrac{1}{4}(\mathbf{a}_{42}-\mathbf{a}_{22}). \end{aligned} \tag{8}$$

COMPARISON WITH EXISTING METHODS FOR TWIST ESTIMATION

Various methods for estimating the quantity $s_{uv}(i,j)$ $(i,j=0,1)$ or 'twist vector' in the matrix S appearing in the Ferguson patch equation (6) have been examined by Barnhill *et al.* (1988). They conclude that the estimation known as 'Adini's twist' gives the best result and consequently suggest its use. For a Ferguson patch within a quadrilateral network of patches, Adini's twist is given by:

$$
\begin{aligned}
\mathbf{s}_{uv}(u_i, v_j) = {} & \frac{\mathbf{s}_v(u_{i+1}, v_j) - \mathbf{s}_v(u_{i-1}, v_j)}{u_{i+1} - u_{i-1}} \\
& + \frac{\mathbf{s}_u(u_i, v_{j+1}) - \mathbf{s}_u(u_i, v_{j-1})}{v_{j+1} - v_{j-1}} \\
& - \frac{\mathbf{s}(u_{i+1}, v_{j+1}) - \mathbf{s}(u_{i-1}, v_{j+1}) - \mathbf{s}(u_{i+1}, v_{j-1}) + \mathbf{s}(u_{i-1}, v_{j-1})}{(v_{j+1} - v_{j-1})(u_{i+1} - u_{i-1})} \qquad (9)
\end{aligned}
$$

(see Fig. 5).

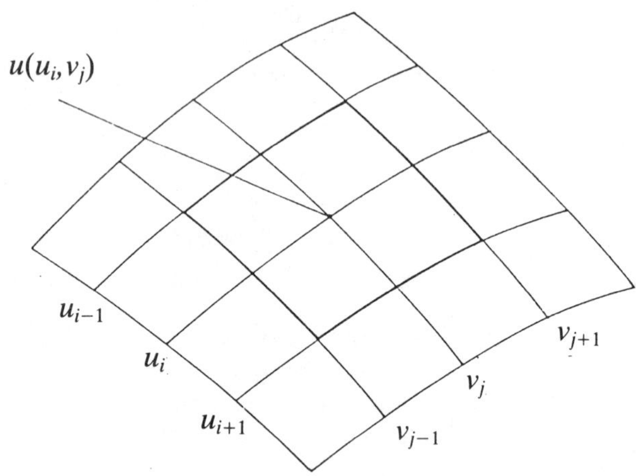

Fig. 5 — Adini's twist.

This quantity is now evaluated for the Overhauser patch, given that the first partial derivatives, $\mathbf{s}_u(i,j)$ and $\mathbf{s}_v(i,j)$, are estimated, using central approximations.

Without loss of generality, consider the corner $\mathbf{a}_{22}$ in Fig. 3, that is, $\mathbf{a}_{22} = s(u_i, v_j)$ and $u_i = 0$, $v_j = 0$. The quantities on the right-hand side of equation (9) are then:

$$
\begin{aligned}
&\mathbf{s}_v(u_{i+1}, v_j) = \tfrac{1}{2}(\mathbf{a}_{33} - \mathbf{a}_{13}) \\
&\mathbf{s}_v(u_{i-1}, v_j) = \tfrac{1}{2}(\mathbf{a}_{31} - \mathbf{a}_{11}) \\
&u_{i+1} - u_{i-1} = 2 \\
&\mathbf{s}_u(u_i, v_{j+1}) = \tfrac{1}{2}(\mathbf{a}_{33} - \mathbf{a}_{31}) \\
&\mathbf{s}_u(u_i, v_{j-1}) = \tfrac{1}{2}(\mathbf{a}_{13} - \mathbf{a}_{11}) \\
&v_{j+1} - v_{j-1} = 2
\end{aligned}
$$

Therefore, after substituting into the right-hand side of equation (9),

$$\begin{aligned} s_{uv}(u_i, v_j) &= \tfrac{1}{4}(\mathbf{a}_{33} - \mathbf{a}_{13} - \mathbf{a}_{31} + \mathbf{a}_{11}) \\ &\quad + \tfrac{1}{4}(\mathbf{a}_{33} - \mathbf{a}_{13} - \mathbf{a}_{31} + \mathbf{a}_{11}) \\ &\quad - \tfrac{1}{4}(\mathbf{a}_{33} - \mathbf{a}_{31} - \mathbf{a}_{13} + \mathbf{a}_{11}) \\ &= \tfrac{1}{4}(\mathbf{a}_{33} - \mathbf{a}_{13}) - \tfrac{1}{4}\mathbf{a}_{31} - \mathbf{a}_{11}) \\ &= s_{uv}(0,0) \text{ in equations (8).} \end{aligned}$$

Henceforth, it may be stated that the Overhauser parabolically blended surface patch is in fact a Ferguson patch with centrally estimated first partial derivatives and Adini's twists.

APPLICATIONS OF OVERHAUSER CURVE SEGMENTS AND PATCHES

The last year at NEI Parsons has seen the development by the authors of a special purpose modeller, particularly suitable for turbine blades.

A turbine blade is defined by a number of parallel cross-sections each containing an equal number of points, or nodes. By joining corresponding nodes on each cross-section, they naturally form a quadrilateral mesh of nodes on the surface, thus making the turbine blade ideal for modelling by piecewise bicubic patches. Moreover, the Overhauser patch is particularly favourable as it is defined by nodes already on the surface, and no interaction is therefore required by the user of the program.

The result is a complete definition of the surface bounding the turbine blade. This feature makes it easy to generate realistic images. Wireframe images (Fig. 6) can be quickly generated by plotting the boundary of each patch, consisting of the isoparametrics $u = 0,1$ and $v = 0,1$. Rudimentary greyscale images can also be displayed fairly rapidly. Each sub-patch is calculated:

$$n = \frac{\mathbf{r}_u(u,v) \times \mathbf{r}_v(u,v)}{\|\mathbf{r}_u(u,v) \times \mathbf{r}_v(u,v)\|} \, .$$

By testing to see how this normal lies with respect to a light source vector, defined by the user, the subpatch can be coloured accordingly. The 'painter's algorithm' is used to perform any necessary hidden surface removal (Newman & Sproull 1981).

A number of utility routines have also been developed. By numerically integrating certain integrands over the surface of the blade, properties such as the volume, centre of volume, and moments of volume are calculated.

A complete breakdown of the section properties required by blade designers is available. Composite Overhauser curve segments define the section boundary. The elementary properties associated with the blade section, such as the area, moments of area, etc., are determined by numerically integrating certain functions around the boundary. More complex quantities, the centre of flexure, for example, are calculated by using the solutions of a number of partial differential equations within the blade section (Sokolnikoff 1956). Because the Overhauser curve segments use only nodal values in their definition, the boundary element method is employed to solve these equations.

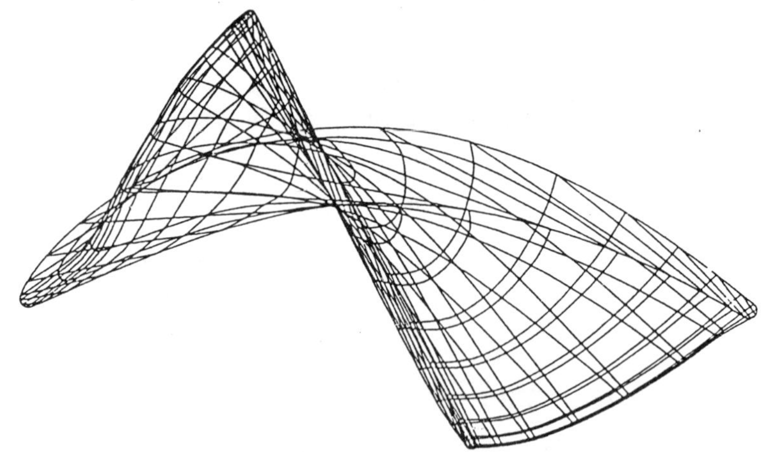

Fig. 6 — Wireframe image of highly-twisted turbine turbine blade.

Finally, the section properties for a number of sections at various heights up the blade have been combined in a routine which determines the amount by which a highly twisted low pressure blade will untwist at running speed. This entails numerically solving a series of ordinary differential equations (Montoya 1964).

ACKNOWLEDGEMENTS

The authors would like to thank the National Advisory Body for financing this project and also NEI Parsons Ltd for access to their computing facilities as well as their financial support.

REFERENCES

Barnhill, R. E., Brown, J. H. & Klucewicz, I. M. (1978) A new twist in computer aided geometric design, *Computer Graphics and Image Processing* **8** 78–91.

Barnhill, R. E., Favin, G., Fayard, L. & Hagen, H. (1988) Twists, curvatures and surface interrogation, *Computer Aided Design* **20** 341–346.

Faux, I. D. & Pratt, M. J. (1979) *Computational geometry for design and manufacture*, Ellis Horwood, Chichester.

Hall, W. S. & Hibbs, T. T. (1987) The treatment of singularities and the application of the Overhauser C^1 continuous quadrilateral boundary element to three dimensional elastostatics, In: *Advanced Boundary Elements*, IUTAM symposium, San Antonio, Texas.

Montoya, J. (1964) Torsion of warped beams subjected to axial stress, *Ingeniera Naval* **349** (July).

Newman, W. M. & Sproull, R. F. (1981) *Principles of interactive computer graphics*, McGraw-Hill, New York.

Overhauser, A. W. (1968) Analytic definition of curves and surfaces by parabolic blending, *Scientific Research Staff Publication*, Ford Motor Company, USA.
Sokolnikoff, I. S. (1956) *Mathematical theory of elasticity*, McGraw-Hill, New York.

14

Surface roughness characterization in fluid flow

John S. Medhurst
Vine Abel & Co. Ltd, Maritime Consultants

The link between surface roughness and resistance to fluid flow is as yet imperfectly understood. This chapter sets out the simplified logarithmic law of the wall and roughness function approach to the fluid mechanics of the problem. Surface roughness is characterized from the statistical properties of profiles digitized from a stylus transducer. Correlations between the hydrodynamic roughness and profile statistics are examined and some practical applications outlined. Roughness measurement is particularly sensitive to method divergence, and standard procedures for profile characterization are therefore required before further progress can be made.

1 INTRODUCTION

The increase of fluid resistance with surface roughness is a complicated and expensive phenomenon. A moderate level of wetted hull roughness may increase a shipowner's fuel bill by 10%, and for a large tanker the fuel cost may be $1m per annum, even at today's artificially low fuel prices. A further 5% can often also be saved by polishing the propeller. Operators of large pipeline projects also suffer high penalties from surface roughness. Injection of long chain polymers at the source and retrieval at destination has even been tried as a means of reducing pumping power in overland pipelines.

Smaller pipework systems are unlikely to gain significant cost benefit from reductions in wall surface roughness, but the designer of the system will still need to know the effect of roughness on resistance so that flow rates can be judged accurately and pumps and pipes can be correctly sized.

However, ship hull resistance, besides being the author's particular research interest, is probably the most tangible manifestation of surface roughness in fluid flow, and it is no coincidence that some of the most significant progress in tackling the problem generally has been made by the maritime community.

It was realized even before the days of steam that barnacle and weed growth and worm damage of wooden ships slowed them down. One early but expensive solution to the problem of marine growth was copper cladding, believed to be the origin of the phrase 'copper bottomed'. Techno-economic studies extolling the virtues of cupro-nickel sheathing make an appearance from time to time. However, the single most significant advance in antifouling technology, and arguably in the whole field of ship maintenance, spurred on by the oil price rises of the early 1970s, was the invention on Tyneside of a self-polishing copolymer antifouling (tributyl tin oxide) which, as its name implies, polishes away as its toxic effect becomes exhausted, leaving a fresh layer of toxin underneath. Standard drydocking intervals lengthened from 12 to 30 months in a few years as these paints became standard. Now they are under threat from environmentalists, but the mould has been broken; shipowners expect ships to remain foul free, and the new tin-free and non-toxic antifoulings now being developed will continue to accomplish this.

With the fouling problem to a large extent conquered, ship researchers began to concentrate with renewed energy on keeping ships smooth as well as clean and monitoring their surface condition at drydocking, before and after painting. The development of a portable stylus gauge, also on Tyneside, the BMT Hull Roughness Analyser or HRA (a logical development of the earlier Wall gauge) allowed marine superintendents and surveyors to do just this. Ten years later it is the only instrument which can measure 100 locations on a ship hull in the single day (maximum) available for access in the drydock before and after painting.

Surface metrology as a science did not develop significantly until the 1960s and 70s. It has now come of age with the publication of an excellent textbook (Thomas 1982). However, in the early days of research into roughness and drag it simply did not exist.

The great 19th century hydrodynamicist W. E. Froude realized the significance of roughness, and as part of his programme of measuring the drag of planks in a towing tank in his garden at Torquay, included coverings of calico and graded sand of different sizes. Later, Nikuradse carried out a consistent series of experiments using a single size of sand grains in pipes of different diameters, work which is quoted in all standard texts on fluid mechanics. This work was extended to flat plate flows by Schlichting (1983). The Nikuradse–Schlichting charts were later modified for engineering type roughness by Colebrook & Moody (pipes) and Hama (flat pipe boundary layers), but the unit of roughness still remained the 'equivalent sand grain diameter', K_e. Finding a correlation between K_e and the properties of a particular surface relied on the intuition of the engineer.

With the development of surface metrology, and particularly the surface profilometer, a start has been made in the last 15 years on finding a link between the statistical parameters of surface profiles and fluid drag. Approximate correlations between standard parameters such as centre line average height (R_a) and K_e are often found in the instruction manuals of proprietry surface gauges, but they should be treated with some caution.

Since this is a book on surface definition the fluid mechanics which follows will be kept to a minimum and will concentrate on introduction of the universal hydrodynamic roughness function and a 'hydrodynamic roughness number'. We will then examine the statistical properties of surfaces which may be of interest and review the state-of-the-art in correlating surface roughness and fluid drag for a number of practical problems. Finaly, we will outline a possible standard for the characterization of surface roughness in fluid flow.

2. FLUID FLOW PAST SMOOTH AND ROUGH SURFACES

We are here concerned solely with the inner part of the boundary layer formed by the flow of a fluid past a solid surface at sufficient velocity to ensure turbulent flow. At the wall (solid surface), the velocity is zero and the flow exerts a shear stress τ_w. In the few tens of μm closest to the wall the turbulence is damped by the presence of the wall. This region is known as the sublayer. Outside the sublayer and extending out to about 10 or 15% of the boundary layer thickness the velocity increases logarithmically with distance from the wall (see Fig. 1):

$$u = A \ln y + B \ . \tag{1}$$

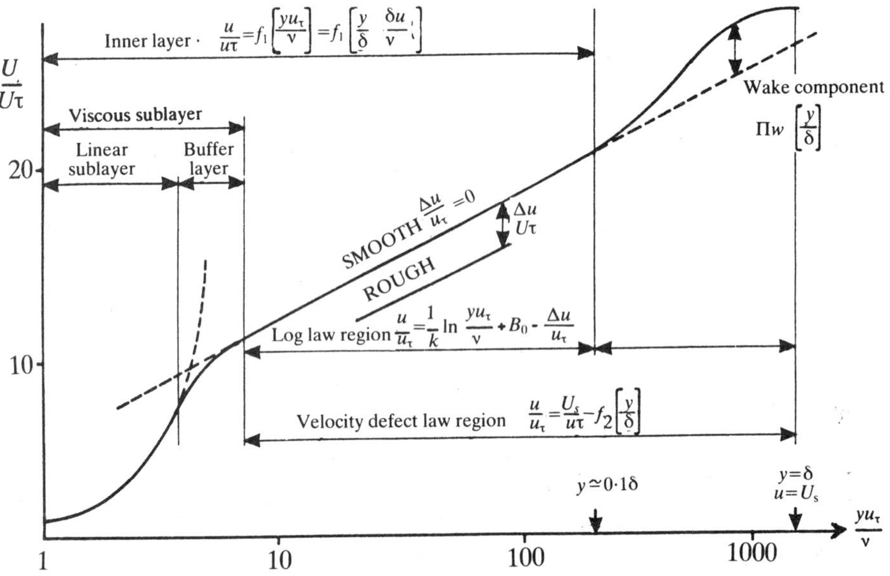

Fig. 1 — The logarithmic law of the wall and roughness function.

For dynamic similarity:

$$\frac{u}{u_\tau} = \frac{1}{\kappa} \ln \frac{y u_\tau}{\nu} + B_0 \tag{2}$$

where $u_\tau = \sqrt{(\tau_w/\rho)}$, ν is kinematic viscosity, and κ and B_0 are 'universal' constants which generally have values of 0.41 and 5.0 respectively. Equation (2) is known as the logarithmic 'law of the wall'.

In the case of flow past a rough surface, the law of the wall still applies but the velocity distribution is displaced downwards by an amount $\Delta u/u_\tau$ (also written Δu_+). The slope remains the same; only the intercept changes. Thus:

$$\frac{u}{u_\tau}=\frac{1}{\kappa}\ln\frac{yu_\tau}{\nu}+B_0-\Delta u_+ \tag{3}$$

For gross roughness which totally disrupts the sublayer the resistance becomes independent of viscosity:

$$\frac{u}{u_\tau}=\frac{1}{\kappa}\ln\frac{y}{h}-B_r \tag{4}$$

where h is an arbitrary measure of roughness which has the dimensions of length and B_r is a constant, the rough wall equivalent of B_0. Combining (3) and (4):

$$\Delta u_+=\ln\left[\frac{C_1h}{B_1}\frac{u_\tau}{\nu}\right] \tag{5}$$

where $B_0 = -(1/\kappa)\ln B_1$ and $B_r = -(1/\kappa)\ln C_1$.

For hydraulically smooth walls:

$$\Delta u_+=0=\frac{1}{\kappa}\ln(1) \tag{6}$$

Colebrook (1938, 1939) conceived the idea that for a wide range of surfaces having random roughness topography, Δu_+ would be asymptotic to the hydraulically smooth and full rough laws respectively. Hence:

$$\Delta u_+=\frac{1}{\kappa}\ln\left[\frac{(C_1h)}{B_1}\frac{u_\tau}{\nu}+1\right] \tag{7}$$

This is now called the 'Colebrook' form of the roughness function, (see Fig. 2). Since B_0 and κ are universal constants, the quantity (C_1h) can be calculated directly from measurements of Δu_+ (assuming they follow the Colebrook form) and may be termed the 'hydrodynamic roughness number'. (C_1h) has the dimensions of length and is an intrinsic property of the surface.

For Nikuradse close packed sand, $C_1 = 0.0309$ where $h = K_s$, the sand grain diameter. Thus:

$$(C_1h)=0.0309\,K_s \tag{8}$$

which relates the hydrodynamic roughness number to the old concept of 'equivalent sand grain diameter' K_e. For the concept of a hydrodynamic roughness number adds nothing new to the concept of equivalent sand grain diameter, but its derivation is far more satisfactory.

Not all surfaces have a Colebrook roughness function. Homogeneous surfaces (such as close-packed sand) tend to have a 'humped' Δu_+ function in which the transition from hydraulically smooth to fully rough behaviour is more sudden. Other geometrically homogeneous surfaces such as 2-D sinusoids and rectangular bars exhibit similar behaviour. However, inhomogeneous surfaces with varied topo-

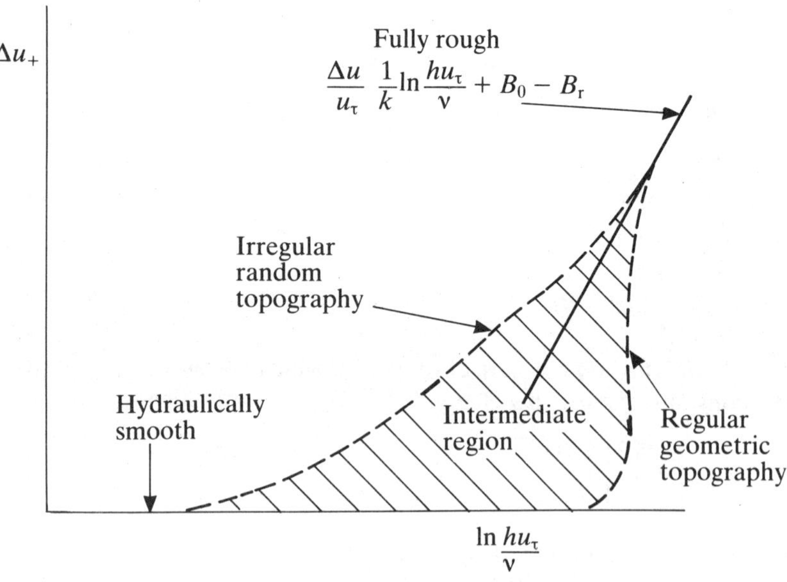

Fig. 2 — The hydrodynamic roughness function.

graphy appear to exhibit Colebrook or similar behaviour. One may model Colebrook surfaces as a 'superposition' of homogeneous surfaces of different roughness amplitude. Fig. 3 shows how 4 homogeneous surface roughness functions may be displaced on the $(C_1h)u_\tau/\nu$ scale to represent variation in amplitude and summed to produce a composite which is very similar to the Colebrook form. This superposition theorem was mooted by Lewkowicz (Lewkowicz & Musker 1978) as a possible means of modelling the Δu_+ function of randomly rough surfaces in terms of their Fourier components. It is one approach to the problem which merits further attention.

The Δu_+ function can be measured in a wide range of laboratory environments including pipes, rectangular ducts, friction planes in towing tanks, wind and water tunnels, and even rotating discs and cylinders. The law of the wall with the Δu_+ function included can then be used as one of a system of equations to calculate the development of the boundary layer on external flows. In the special case of internal flows, the law of the wall itself (with one extra term) is the only equation needed for a solution.

3 MEASUREMENT AND CHARACTERIZATION OF SURFACE ROUGHNESS

Most serious attempts to correlate surface roughness and fluid drag have not relied on sophisticated models such as the superposition theorem but have merely attempted to choose commonly available statistical parameters of individual surface profiles which, when substituted for h in equation (7), cause Δu_+ data to collapse onto a single curve with a suitably chosen value of C_1.

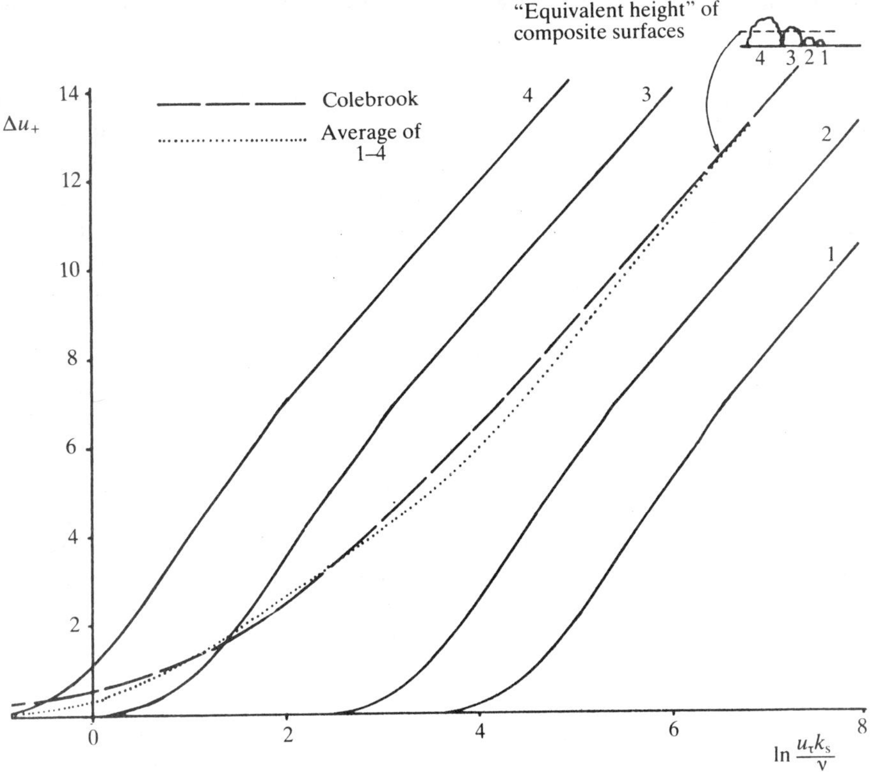

Fig. 3 — Illustration of roughness function superposition.

Whilst a range of standardized methods has been developed for measuring the Δu_+ function there is no standard method for characterizing surface roughness in fluid flow. On the contrary, the subject is bedevilled by lack of standardization, and each research establishment which has not previously, contributed seems determined to go its own way.

The only parameter which approaches being a standard is BMT's Rt_{50} for ship hulls, maximum peak to trough height in a 50 mm cutoff length, as measured by the HRA instrument.

Roughness characterization parameters can be divided into two convenient categories: those which measure roughness height and those which measure the spatial separation of asperities, generally termed 'texture' parameters. The height parameters most commonly in use are CLA (R_a), rms roughness (R_q) calculated either direct or as $\sqrt{m_0}$ where m_0 is the zeroth moment of the Power Spectral Density Function (PSDF), peak to trough height Rt and ten point height R_z. It has been said that event counting parameters such as Rt and R_z ought to be more hydrodynamically significant then R_a and R_q but in fact R_q probably contains most information about the surface condition. For a random Gaussian surface profile, $R_a = \sqrt{2/\pi}R_q$

($\approx 0.8\,R_q$) but the ratio can be as low as 0.5 for lapped and honed bearing raceways (Thomas 1982). A low value of R_a/R_q suggests unusually prominent peaks and/or valleys.

There are numerous candidates for use as hydrodynamic texture parameters. One of the favourites is average slope Sl or the standard deviation of the slope distribution σ_{Sl}. For a Gaussian surface $\sigma_{Sl} = \sqrt{m_2}$ where m_2 is the second moment of the PSDF. Average wavelength λ_a has also been suggested where $\lambda_a = 2\pi R_a/Sl$. Arguably a better parameter (and certainly a proven one) is peak count wavelength λ_{pc}, which is defined as the sample length divided by the number of complete crossings of an envelope of amplitude R_A evenly disposed about the profile centreline.

Correlation length β^* has also been used as a texture parameter. β^* is defined as the lag length over which the autocorrelation function $\rho(\tau)$ (normalized Autocovariance Function ACVF) decays to some arbitrary value, which should be kept as high as possible and is usually fixed at 0.5. β^* is probably not the best choice of texture parameter. Far more information can be obtained from the ACVF by computing the zeroth, second, and fourth derivatives at the origin to estimate m_0, m_2, and m_4.

A combination of these three even moments of the PSDF was suggested by Nayak as a viable texture parameter in surface definition work, the so-called 'Bandwidth parameter' α:

$$\alpha = \frac{m_0 m_4}{m_2^2} \quad . \tag{9}$$

α has been shown to be a useful texture parameter in fluid flow work but is highly susceptible to method divergence, being sensitive to variations in both long and short wavelength cutoffs.

There are alternatives to sine wave or 'Euler' Fourier analysis as means of decomposing random signals into frequency components. One alternative is Walsh analysis (Beauchamp 1975) which decomposes the signal into an ordered set of rectangular waves. Walsh analysis is much faster in computation time than Euler Fourier analysis. The moments of the Walsh PSDF, μ_0, μ_2, and μ_4 may be computed as exact counterparts of the Euler Fourier equivalents, with a corresponding bandwidth parameter β.

A least squares correlation line has been computed but the data is probably better represented by a higher negative slope, for example:

$$1/\lambda_{pc} = -5.0(1 - 1/\alpha) + 5.4 \quad .$$

It is generally agreed that one height and one texture parameter are probably sufficient to achieve a working correlation between surface roughness and fluid drag. Most of the commonly used texture parameters appear to be related. For example α and λ_{pc} are closely related (see Fig. 4) as are α and its Walsh equivalent β (Fig. 5). This may have implications for the superposition theorem outlined earlier since there is far more information available on the Δu_+ functions of rectangular geometric roughness than there is on sinusoidal surfaces.

A correlation has also been established between λ_{pc} and β^*:

$$\lambda_{pc} = 8.8\beta^*_{0.5} \tag{10}$$

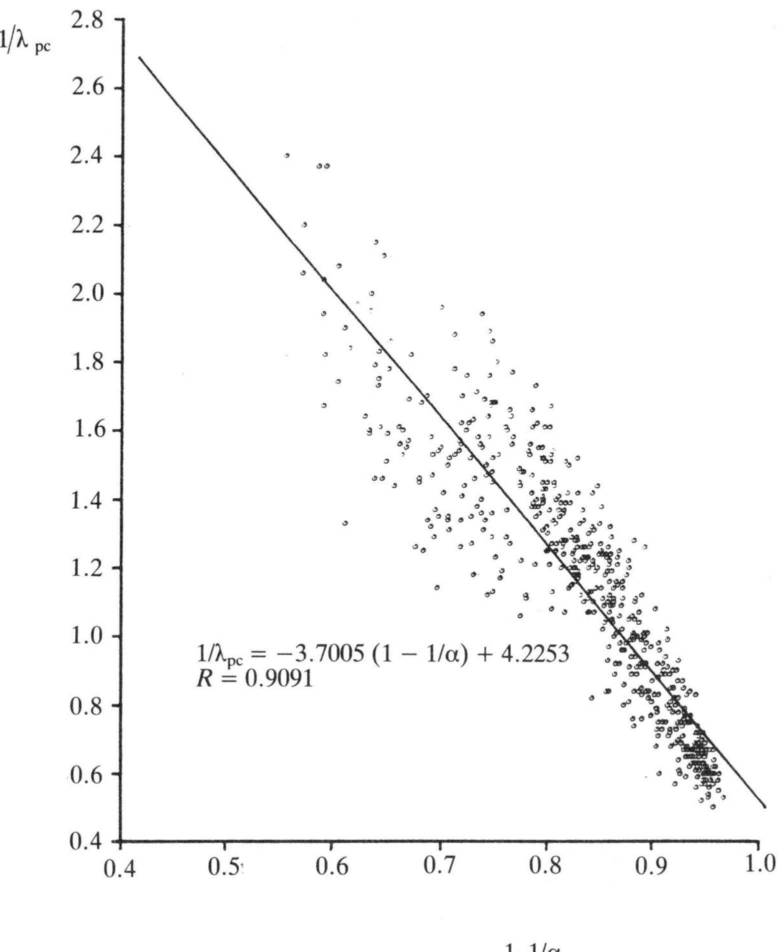

Fig. 4 — Correlation between λ_{pc} and α.

The reason why no satisfactory universal correlation between a combined roughness/texture parameter and fluid drag has yet been found is the lack of standardization and method divergence in profile characterization, making it very difficult to compare the results of work carried out in different research establishments.

4 CUTOFFS AND FILTERS

There is not even any universal agreement on the long and short wavelength cutoffs appropriate to fluid flow work.

The 'standard' parameter for ship hull work is Rt_{50} which is measured at a long

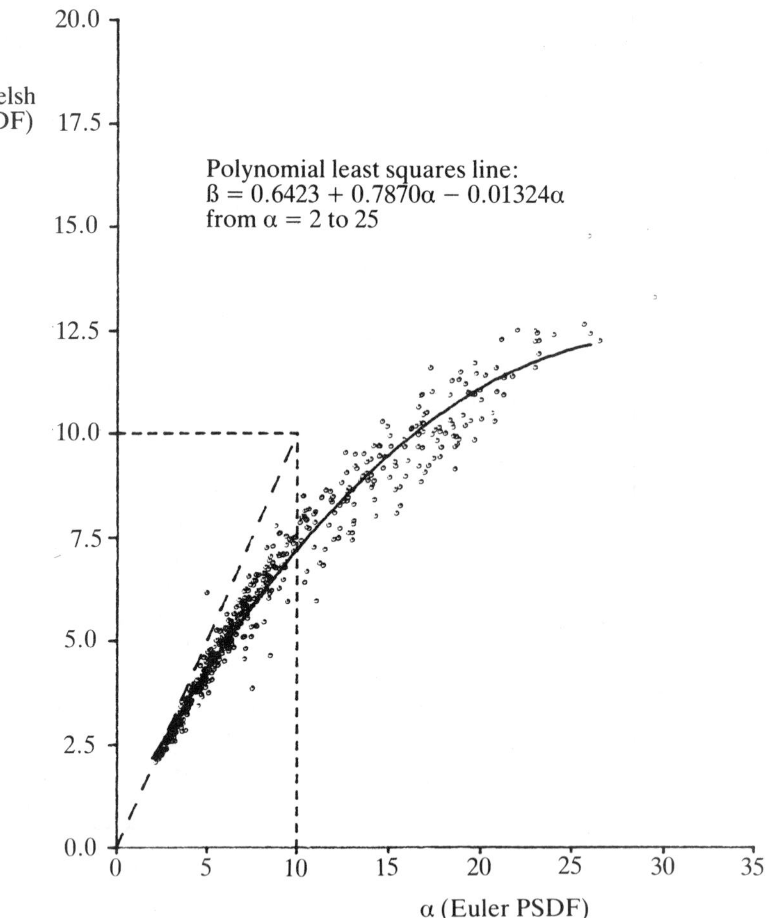

Fig. 5 — Correlation between α and its Walsh equivalent β.

wavelength cutoff of 50 mm. Whilst this parameter is good for assessing the general quality of hull maintenance it is defective for calculating the effect on fluid flow, and tends to be dominated by long wavelengths and high amplitude components, probably caused by the rheology of the paint as its flows during drying and having no effect at all on resistance to fluid flow.

More recently 2 mm has been suggested as a long wavelength cutoff (Musker 1977) but was chosen in simple preference to 50 mm or 10 mm and is almost certainly too short for painted or even machined surfaces. Others have suggested 5 mm (Byrne *et al.* 1982), and this is the author's personal preference of the available ISO cutoffs.

At the short wavelength end of the spectrum, the high pass cutoff for painted surfaces has tended to be dominated by the size of stylus used. The Nyquist criterion suggests that wavelengths shorter than about the tip radius of the stylus will be attenuated. Portable gauges for use in ship hull work must exert a low stylus pressure so that they do not deform freshly dried paint, and they also need to be used at any angle, even upside down on the flat bottoms of merchant ships. Thus both the Wall Gauge and the HRA have ball styli of about 1.6 mm diameter which fixes the high pass cutoff.

Portable gauges used for most other types of surface such as the Rank Taylor Hobson 'Surtronic' and laboratory stylus profilometers do not have this problem. In the case of bench profilometers interfaced to computers (now standard) the digitizing interval can be chosen by the user down to the stylus tip dimensions, typically $3 \times 8\,\mu$m.

There are numerous ways of looking at this cutoff definition problem, including measurement of the size of turbulent eddies and simple trial and error to see which cutoffs give the best correlation with fluid drag. The author has suggested that the cumulative moments of the PSDF may be used as an alternative method of gaining an insight (Fig. 6). The high wavenumber at which m_0' levels out marks a limit beyond

SURFPACK RUN NO: 41
PROFILE I/D: B2CA VENUE: 30–3–84 PROFILE NO: 1
Cumulative Integrals of Moments $m_0, m_2 m_4$ and Bandwidth Parameter α. $m_n' = \int_0^\gamma \gamma^n G_\theta(\gamma)\delta\gamma$

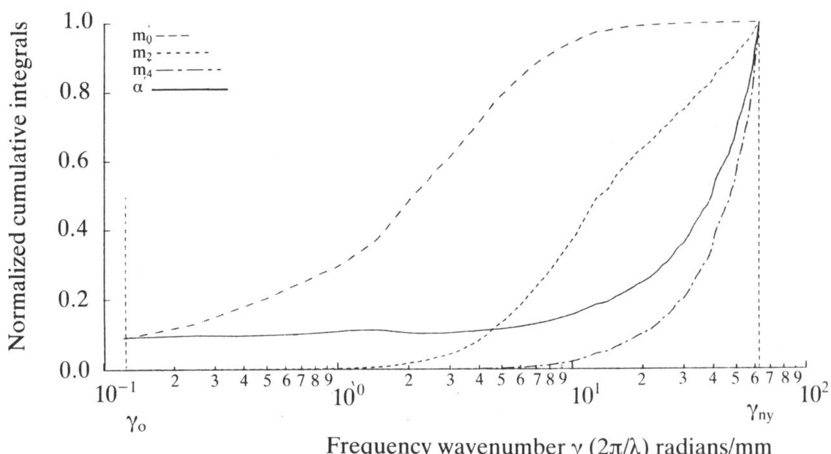

Fig. 6 — Cumulative moments of the PSDF.

which the PSDF gains no appreciable further power and surface features become too small to affect fluid drag. At the low wavenumber end, the point at which m_2' first begins to show a significant value marks the lowest wavenumber which is likely to disrupt the flow.

The best overall combination of long wavelength cutoff and digitizing interval will probably prove to be 5 mm and 100 μm, which gives a wavenumber ratio of 25, arguably sufficient for the application. This choice of digitizing interval would require improvements in the technology of hand held profilometers for painted surfaces.

Whatever the choice of long wavelength cutoff, the profile must be efficiently filtered. If a mechanical filter (skid) or electrical filter (2CR network) is used and is not tuned to the choice of cutoff, filtering by fitting individual centrelines through each cutoff length is not adequate. A simple moving average filter with a window width equal to 80% of the nominal cutoff is an order of magnitude more efficient (contrast Figs 7 and 8 which show filtered and rejected profiles and their respective PSDFs). Other more advances filters, such as the cascaded Z-transform recursive filter described by Beauchamp & Yuen (1979) are better still. Failure to filter correctly leads to method divergence.

5 METHOD DIVERGENCE

Most characterization parameters are sensitive to method divergence, hence the need for standards. By way of example, average slope may be calculated by 2, 3, 5, or 7-point numerical differentiation formulae. All four methods give substantially different values. Incidentally, the simple 2-point method gives the best correlation between σ_{Sl}^2 and m_2 calculated direct from the PSDF.

As a rule, inaccuracies in long wavelength cutoff and/or poor filtering affect ‘height’ parameters, and inaccuracies in digitizing interval affect ‘texture’ parameters. α and the Walsh parameter β are affected by variation at both ends of the spectrum. λ_{pc} and β^*, on the other hand, are affected little by variations in short wavelength cutoff (although they do increase at longer wavelength cutoffs). Thus although they may not be the best performers for fluid flow work they are probably the most robust.

6 CORRELATIONS BETWEEN HYDRODYNAMIC ROUGHNESS AND SURFACE CHARACTERIZATION PARAMETERS

Having measured (C_1h) in a suitable fluid mechanics experiment and measured various characterization parameters in the surface metrology laboratory, the next stage is to seek a correlation between (C_1h) and any parameters which may appear to have physical significance. We should not necessarily expect such a correlation to pass through the origin, since surface parameters will always have a finite value, even when the surface is hydraulically smooth.

In the case of ship hulls, accepting the reality that in the short term we are constrained to use the results from the HRA instrument, we look for a correlation between (C_1h) and Rt_{50}. There is one, but Fig. 9 shows that it is poor. It has nonetheless formed the basis for an approximate formula (see next section).

We can do better simply be reducing the cutoff from 50 mm to 2.5 mm (see Fig. 10). The latter is not an ideal cutoff, but to achieve any results at all using work done by different reserch establishments, one has to accept whatever is available. The ‘rogue’ point shown (not included in the correlation) is for a poor quality flame sprayed surface with individual surface features larger than 2.5 mm in wavelength, and demonstrates the dangers of setting the cutoff length too short.

Turning now to multiparameter correlations, these are still in their infancy, and

SURFPACK90 Run No. 156

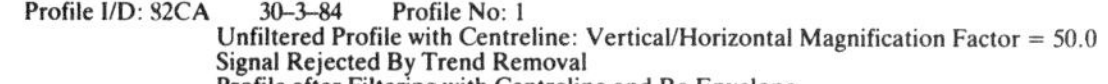

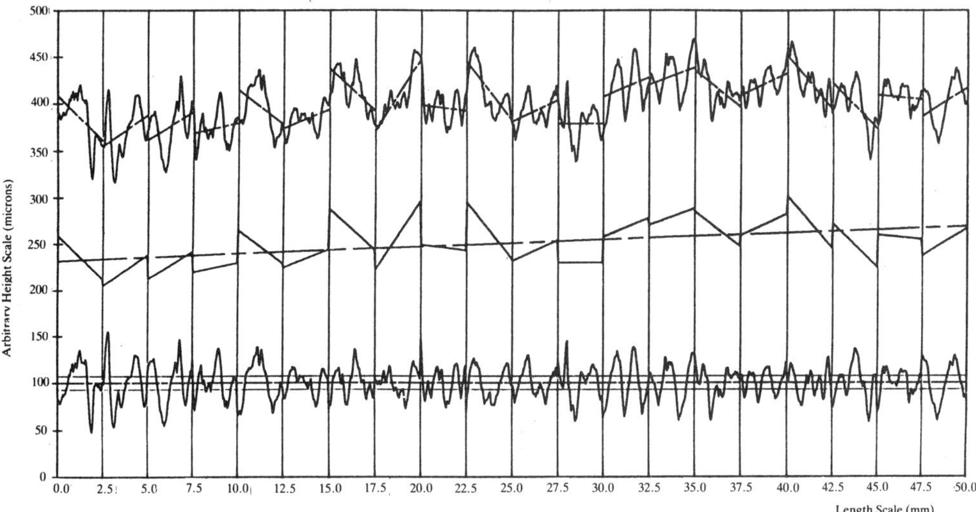

POWER SPECTRAL DENSITY FUNCTION (P.S.D.F.) – SMOOTHED FFT PERIODOGRAM.

Sampling Bandwidth Be = 50.00μm. 18 Degrees of Freedom. 1000 points +24 added zeros.

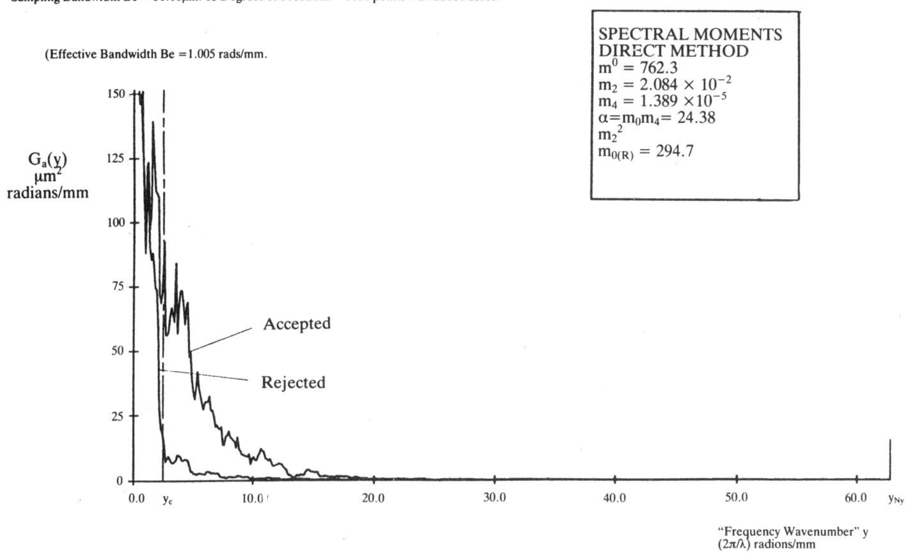

Fig. 7 — Filtering by trend removal (centreline fitting).

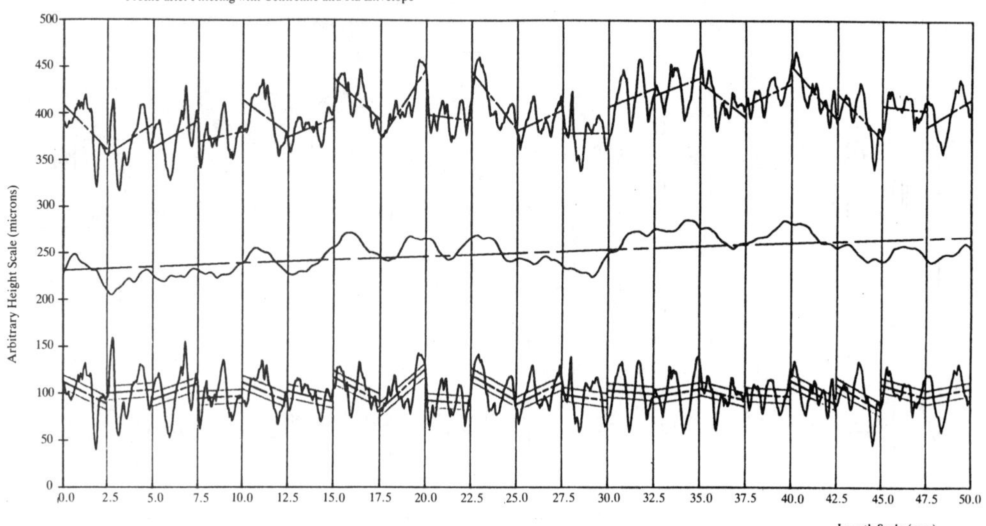

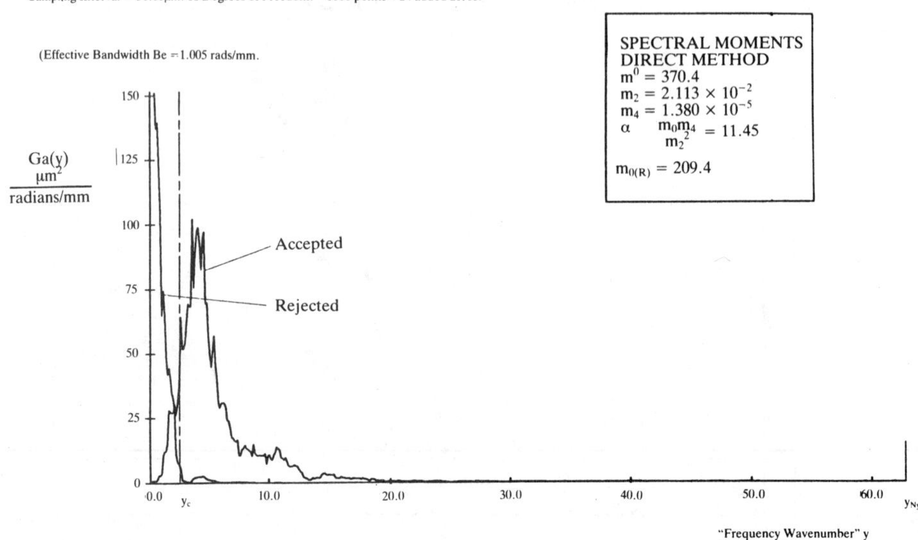

Fig. 8 — Filtering by simple moving average.

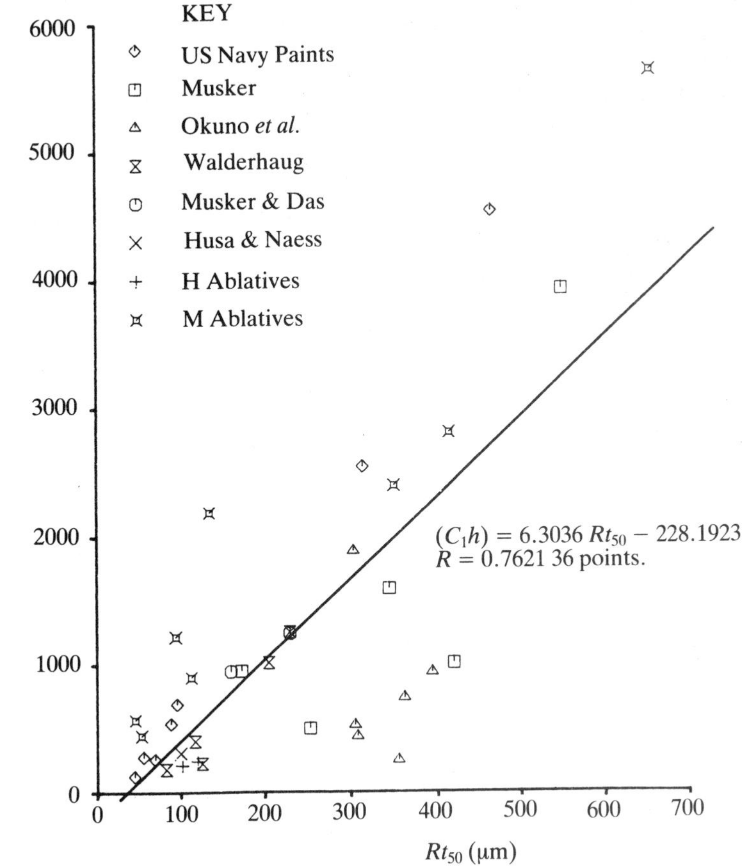

Fig. 9 — Correlation between (C_1h) and Rt_{50}.

the Ship Performance Group at Newcastle University are engaged on the thankless task of trying to bring together work done at various establishments.

One highly convoluted combined height and texture parameter was proposed 10 years ago by Musker (1977) at Liverpool University:

$$h' = R_a(1 + 05\mathrm{Sl})(1 + 0.2kK_u) \tag{11}$$

for which $C_1 \approx 0.02$.

A long wavelength cutoff of 2 mm and a digitizing interval of 50 μm were used. The slope was based on a 7-point Stirling formula with a sampling interval equal to the correlation length. In fact the slope at a fixed digitizing interval of 50 μm could have been used with little loss of accuracy, in which case the constant in the *Sl* term becomes 0.3.

This complicated parameter has not found universal favour, mainly owing to failure of the S_k/K_u term. More recently the Ship Performance Group suggested an alternative based on the same data (Townsin *et al.* 1985).

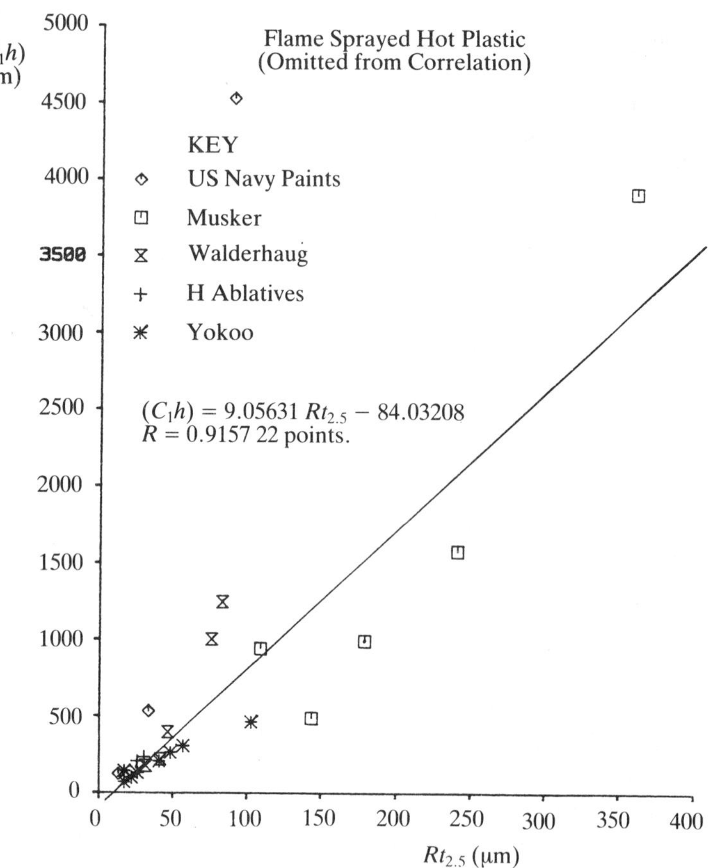

Fig. 10 — Correlation between (C_1h) and $Rt_{2.5}$.

$$h' = 53\frac{\mathbf{R}_a^2}{\beta_{0.5}^*} \tag{12}$$

at a cutoff of 2.5 mm and a digitizing interval of 50 μm. This appears to have wider application but is not universal. The author has also proposed (Medhurst 1989):

$$h' = R_a(0.55\alpha - 1.12) \tag{13}$$

at the same cutoffs, based on a reanalysis of the Liverpool data by Byrne (1980). $C_1 \approx 0.02$ throughout.

None of these combined parameters are perfect, but they are probably better then relying on roughness height alone.

7 APPROXIMATE FORMULAE FOR PRACTICAL APPLICATIONS

For ship hulls, the use of Fig. 9 in a boundary layer integral prediction method leads to:

$$\Delta C_F = 44\left[\left(\frac{\text{AHR}}{\text{Lpp}}\right)^{1/3} - 10R_n^{-1/3}\right] + 0.125 \qquad (14)$$

for evenly distributed hull roughness (AHR) between 100 and 600 μm.

For pipes, Colebrook's famed resistance law, which was the basis of Moody's chart, may be restated:

$$f^{1/2} = -4\,\text{Log}_{10}\left[\frac{(C_1h)}{0.115d} + \frac{1.26}{R_e f^{1/2}}\right] \qquad (15)$$

where f is the English version of the pipe friction factor,

$$f = \tau_w/\tfrac{1}{2}\rho\bar{u}^2 \ . \qquad (16)$$

For rectangular ducts and flumes, equation (15) can be used as a first approximation, provided that d is substituted by mean hydraulic diameter d_m:

$$d_m = \frac{4 \times \text{cross sectional area}}{\text{wetted perimeter}} \qquad (17)$$

For flat plates in a zero pressure gradient boundary layer it is also possible to arrive at one line formula for both local and overall skin friction, but these are intractable. Charts for local flat plate skin friction appear in the literature. Good examples will be found in Hama (1954) and Musker (1977).

For marine propellers, Svenson and the author showed some years ago that the 'one third power law' of equation (14) was also applicable to propeller blades. For the particular propeller studied, the added power requirement due to surface roughness was:

$$\frac{\Delta P}{P} \times 100\% = 1.107(50(C_1h))^{1/3} - 1.749 \qquad (18)$$

This work was later extended to a wider range of propellers by Townsin *et al.* (1985). The method included a monograph and is intended for use by practitioners, but is too lengthy to set out here.

8 THE NEED FOR STANDARDS

To make progress with the hydrodynamic roughness problem, an International Standard is therefore urgently required for the measurement and characterization of surface roughness records in fluid flow work. The standard should cover the following topics:

(1) Long and short wavelength cutoffs. Where possible profiles should be characterized at a range of cutoffs. However, standard cutoffs should be agreed upon at the earliest opportunity. The author is in favour of a long wavelength cutoff of 5 mm and a short wavelength cutoff of 200 μm (digitizing interval 100 μm).

(2) The standard high pass filter should be agreed upon. The author's first choice

would be the cascaded infinite impulse response (IIR) filter but to avoid method divergence, the simple 'boxcar' moving average would be a safe alternative, with the window width set equal to 80% of the nominal cutoff. Thus for the cutoffs suggested above a 41 point moving average would be used.

(3) The length of profile trace should in every case by stated. It is suggested that where possible, multiples of 2050 (plus filter window) should be used. After normal statistical characterization, 2 points per 2050 can be omitted (at the beginning and end of the profile) for calculation of spectral estimates, using a conventional 2^n space-decimation FFT. The need to add zeros can then be avoided.

(4) Methods for calculating the various parameters should be standardized. The method of numerical differentiation for slopes, curvatures, and the derivatives of the ACVF at the origin is particularly important. It is suggested that first difference formulae should be used in every case, with the exception of the ACVF derivatives which should be calculated by the 5 and 7 point Stirling formula for the 2nd and 4th derivatives at the origin respectively.

(5) The practice of fitting a separate least squares line through each cutoff length is unnecessary and should be discontinued. The zero datum provided by the filter method will be adequate for normal parameter characterization. However, a separate single least squares centreline should be fitted though the whole profile trace after filtering to eliminate any residual d.c. level. All parameters should then be calculated continuously over the whole filtered profile length with the exception of R_t and R_z, for which the profile can be divided into nominal cutoff lengths.

(6) A standard method for calculating the Euler and Walsh PSDFs should be stipulated. The method outlined in Appendix 4.1 of Medhurst (1989), based on Newland (1980) is believed to be as good as any. The number of χ_k^2 degrees of freedom should clearly be stated on all PSDF presentations. In the author's view, use of a cosine taper window for surface profile characterization is not usually necessary.

(7) When presenting ensemble averaged parameters from the same surface, the following statistical parameters should be listed:

- Mean
- Standard deviation
- Modal value (approximate)
- The number of profiles measured and the area of the test panel.

(8) Finally, a standard range of parameters should be agreed upon. It is suggested that these should be initially R_t, R_z, R_a and R_q, together with Sl, Cp, Cv, $\beta^*_{0.5}$, λ_{PC}, S_k, K_u, D_e, D_z, and the Euler spectral moments m_0, m_2, m_4 and if possible their Walsh equivalents.

These simple rules are all that are required to establish good working practices which will enable results measured in different laboratories to be compared and correlated with confidence, thus leading to a better understanding of the link between surface roughness and fluid drag.

REFERENCES

Beauchamp, K. G. (1975) Walsh functions and their applications. Academic Press, ISBN 0-12-08450-2.

Beauchamp, K. G. & Yuen, C. K. (1979) *Digital methods for signal analysis*. Allen & Unwin, ISBN 0-04-621027-X.

Byrne, D. (1980) The hull roughness of ships in service. MSc thesis, Newcastle University, October.

Byrne, D., Fitzsimmons, P. A. & Brook, A. K. (1982) Maintaining propeller smoothness: a cost effective means of energy saving. *SNAME Ship Costs and Energy* '82 *Symposium*, *October* 1982.

Colebrook, C. F. (1938–1939) Turbulent flow in pipes, with particular reference to the transition region between the smooth and rough pipe laws. *Jou. Inst. Civil Engs*. **11** 133–156.

Hama, F. R. (1954) Boundary-layer characteristics for smooth and rough surfaces. *Trans. S.N.A.M.E.* **62** 333–358.

Lewkowicz, A. K. & Musker, A. J. (1978) The surface roughness and turbulent wall friction on ship-hulls: interaction in the viscous sublayer. *Int. Symp. on Viscous Ship Resistance*, *Gothenberg*, Paper No. 10.

Medhurst, J. S. (1989) The systematic measurement and correlations of the frictional resistance and topography of ship hull coatings, with particular reference to ablative antifoulings. PhD thesis, University of Newcastle-upon-Tyne, August.

Moody, L. F. (1944) Friction factors for pipe flow. *Proc. A.S.M.E.* **66** 671–684.

Musker, A. J. (1977) Turbulent shear-flows near irregularly rough surfaces with particular reference to ships hulls. PhD thesis, University of Liverpool, December.

Newland, J. (1980) *Random vibrations and spectral analysis*. Longman, ISBN 0-582-46335-1.

Schlicting, H. (1983) *Boundary layer theory*. McGraw-Hill (7th edition) ISBN 0-07-055334-3.

Svensen, T. E. & Medhurst, J. S. (1984) A simplified method for the assessment of propeller roughness penalties. *Marine Technology* **21** (1) January.

Thomas, T. R. (ed.) (1982) Rough surfaces. Longman, ISBN 0-582-46816-7.

Townsin, R. L., Medhurst, J. S., Hamlin, N. A. & Sedat, B. S. (1984) Progress in calculating the resistance of ships with homogeneous or distributed roughness. *NECIES Centenary Conf. on Marine Propulsion*, *Newcastle*, *May* 1984.

Townsin, R. L., Spenser, D. S., Mosaad, M. & Patience, G. (1985) Rough propeller penalties. *SNAME Autumn Meeting*, 1985.

15

Statistics and surfaces — a case study

C. Craggs
Department of Mathematics & Statistics, Newcastle Polytechnic
D. J. Gardiner
Department of Chemical & Life Sciences, Newcastle Polytechnic

INTRODUCTION

Over the past few years, one aspect of our research has been concerned with spatially located or spatially related data. This work has been of a very applied nature as it has usually been attached to a research project, for example in chemistry. Frequently there are conflicts between the requirements made by the chemist and the mathematical/statistical aspects. For example, the chemist has limited time and resources to collect information or is restricted by the instrumentation, whereas a statistician/mathematician would have preferred more data or a different type of sampling strategy. The chemist usually requests that the observations are represented in some way — a 3-dimensional plot or contour lines — and initially frequently does not recognize the importance of model fitting. The case study presented here illustrates some of the practical difficulties encountered.

BACKGROUND

Using a Raman microspectroscope, observations in the form of spectra were collected from locations in a region of approximately 20×20 μm on a surface. Initially, white light is used to obtain a photographic image of the region. A beam of laser light is then focused onto particular locations of this region, and a spectrum of vibrational frequencies is obtained at each of these locations. These vibrational frequencies can be likened to a thumb print as they are unique for each molecular species.

Fig. 1 shows a Raman band. The area under a particular Raman band is related to the concentration of a molecular species, and a movement in a Raman band is related to the stress in a surface or pressure in a liquid. In this particular study, the latter

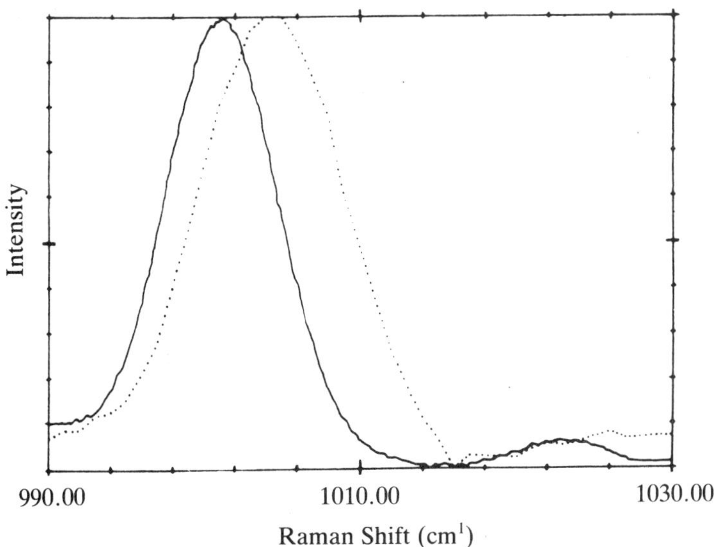

Fig. 1 — Raman band of polyphenyl ether (lubricant) showing shift due to difference in pressure within the ball/plate contact (.........) and outside the ball/plate contact (—).

feature was considered, and the movement in the Raman band in all datasets presented here was measured visually by the chemist.

ELASTOHYDRODYNAMIC CONTACTS

In high-load bearing contacts, the bearing surfaces elastically deform around a thin layer of lubricant. These elastohydrodynamic contacts can be modelled, using a steel ball which is covered by a thin film of lubricant rolling on a glass plate (Fig. 2). Using Raman spectroscopy, the spectrum of the lubricant at points in the contact region can be observed (Baird 1988 for full details). The relative pressure experienced at a point in the contact was assessed by observing the change in frequency of a particular Raman band. This change was recorded such that a high value indicates a high pressure. (The system was designed as a 100% rolling contact so that no shear was present.)

A typical dataset was analyzed here. The applied load was 7.2 kg and the contact was rolling at a speed of 0.104 m s^{-1}. There are 30 pressure observations (kbar) on the contact and 11 zero observations which indicate the edge of the contact (as the pressure on the edge and outside of the contact is theoretically zero). The locations are given in Fig. 3. Originally, these data were collected to construct profiles, and so observations were taken on only half of the surface. To obtain a more realistic representation, it was assumed that the 'lower' (i.e. negative) half of the contact was a mirror image of the 'upper' half.

Using least squares estimation, an ellipse, originally suggested by the chemists,

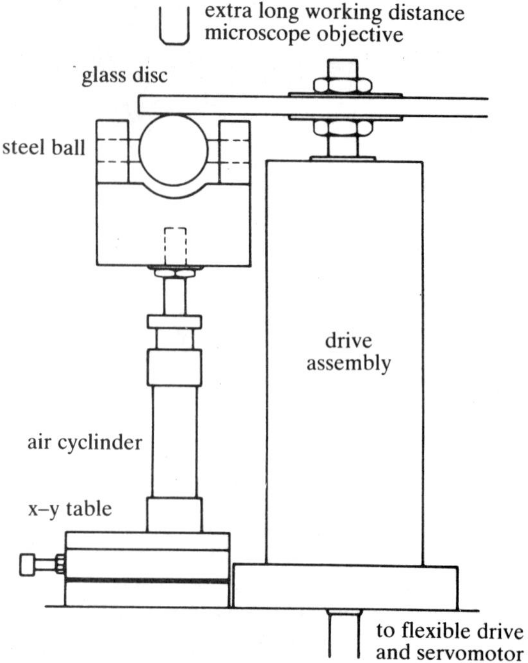

Fig. 2 — Dynamic ball/plate Raman lubrication rig.

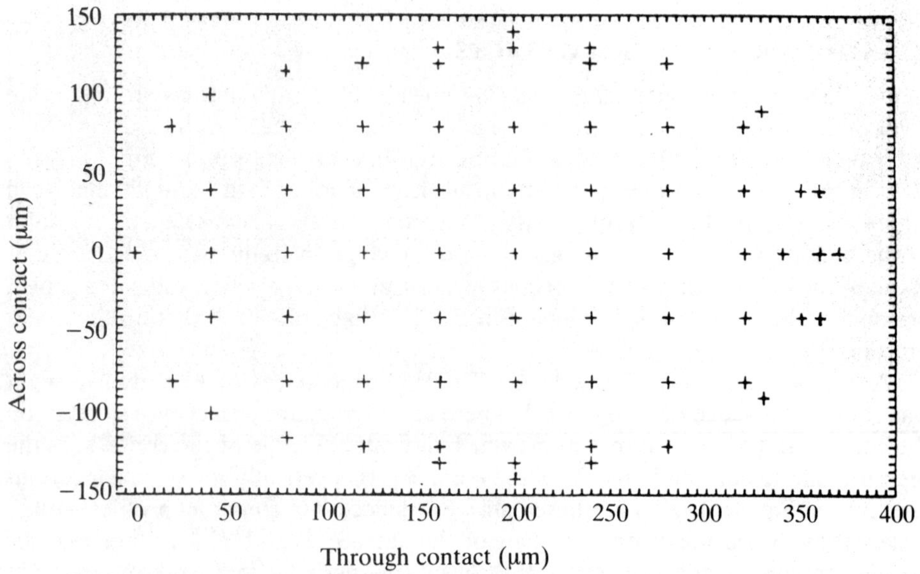

Fig. 3 — Locations of observations from an elastohydrodynamic contact.

was found to be a good model for the outline. A set of dummy zero values laying on this ellipse was then used to indicate the edge of the contact.

THE PROBLEM

The research workers required the observed data to be represented in the form of 3-dimensional plots and contour maps for a number of datasets.

At this stage in the development of the instrumentation, the available data in each dataset were extremely limited as it took from 5 to 30 minutes to obtain a single spectrum. It was therefore important to model the available data carefully and consider the position of sampling points for future experiments.

MODELLING

Choice of modelling package was dependent on a number of factors: ability to produce good quality graphics (3-D plots and contour maps), availability, flexibility, controllability, and the capacity to cope with data taken at 'random' points. (This latter factor was for other related datasets.) At this time, SURFACE2 was the only available package which was able to do this. The user can specify a wide range of weights and smoothing functions. In addition, the residuals and predicted values can be obtained and used for further analysis. It should be noted that all 3-dimensional plots and contour plots given in this paper are as obtained from SURFACE2.

There are two methods of estimation available in SURFACE2. In the first method, each sample data point is estimated by a distance weighted average of the sample data points found in the specified search area around that grid intersection. This method was used to obtain the results shown here.

In the second method, a first order trend surface is fitted at each sample data point, using nearby sample data points. The trend surface is constrained so that it passes through the control point being evaluated, and so it is a least squares estimation of the dip of the surface at that point. Coefficients of this trend surface are then evaluated for the grid node being estimated — in effect projecting the dip of the surface from the sample data point to that grid intersection. A distance weighted average is then made of the projected dips from the sample data points found in the specified search area around the grid point.

RESULTS

Five different weighting factors are available (Table 1). Each was used to model the dataset, and a residual analysis was carried out. Results (Table 1) show that, in this particular case, the weighting factor is optimized with a value D^{-6}.

A contour map and 3-dimensional plot for the pressure are given in Figs 4 and 5 respectively. It can be seen that the pressure is generally as expected from theoretical models. A small number of 'spikes' on the edge are thought to be due to the relatively small number of observations and the complex situation being modelled. The weighting factor D^{-6} was thought to be optimal owing to the sharp drop in levels at the exit edge of the contact.

Table 1 — Different weighting factors used to fit dataset given in Fig. 3, using SURFACE2 and corresponding sum of squared errors

Weight	Sum of squared errors
$(1-D/(1.1^*D_{max}))^2/(D/(1.1^*D_{max}))^2$	0.01006
D^{-1}	2.49769
D^{-2}	0.06780
D^{-4}	0.00033
D^{-6}	0.00001

where D_{max} is the distance from the grid intersection to the most distant sample point in the set being used in the estimation

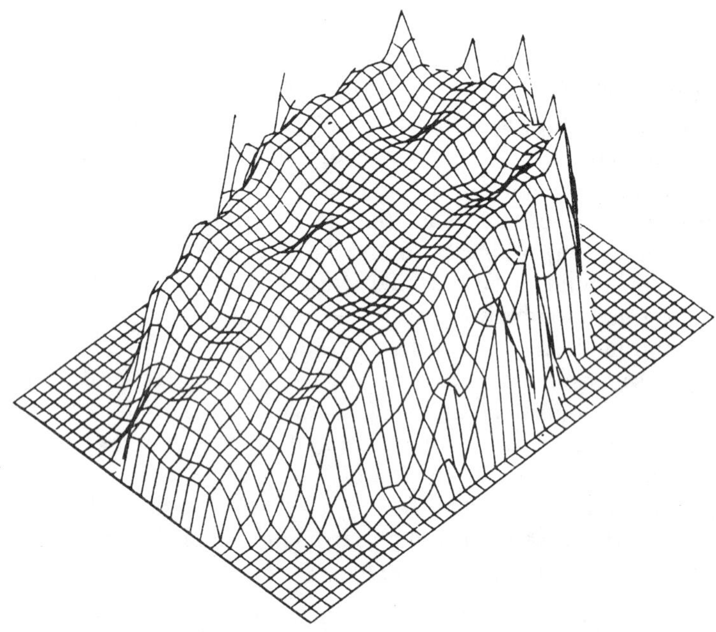

Fig. 4 — Three-dimensional plot of dataset in Fig. 3 obtained from SURFACE2 with weighting factor D^{-6}.

SIMULATION

At this time, it was possible to collect only a very limited number of observations in each experiment. The sampling strategy was therefore very important, the different strategies were investigated by simulating a surface.

The simple situation of the relative stress values on a grain of oxide was

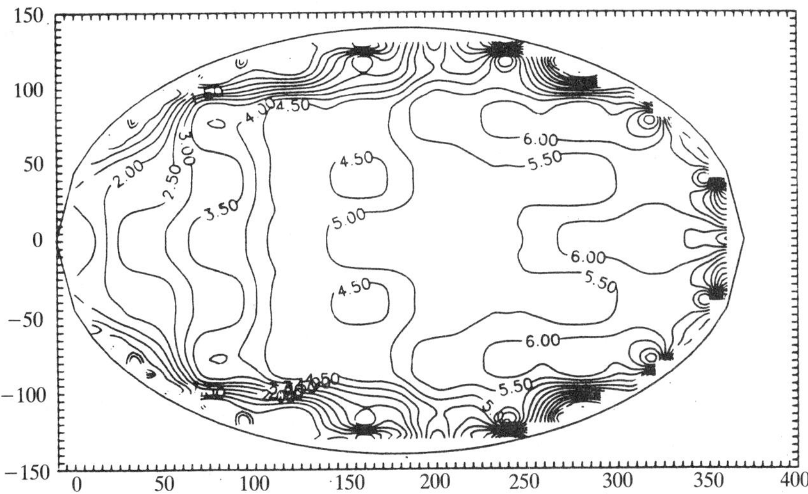

Fig. 5 — Contour plot of dataset in Fig. 3 obtained by using SURFACE2 with weighting factor D^{-6}.

simulated. In experimental work, it was found that the stress on the grain was considerably higher than on the boundary. Unfortunately, as the available datasets were very small, the noise distribution was difficult to assess. However, a normal distribution appeared to be a reasonable approximation. For the simulated surface, a mean of 4 and standard deviation of 1 was used for the grain, and a mean of 7 and standard deviation of 1 was used for the boundary. Fig. 6 shows a 3-dimensional plot

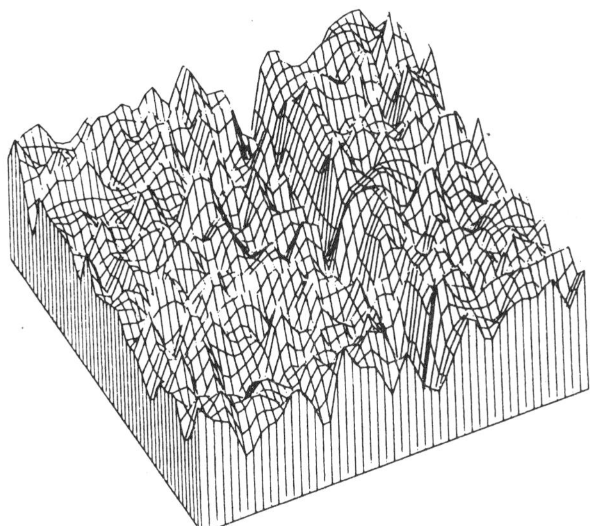

Fig. 6 — Three-dimensional plot of a simulated surface.

of such a surface. The situation simulated is a 25×25 grid of points (i.e. 625 observations). There are two rows which represent the boundary and can clearly be seen as a dip.

The conditions under which the dip is observable is being investigated by (i) changing the relative values of the mean and standard deviations, and (ii) reducing the number of observations taken on the surface.

FACTORS INFLUENCING RESULTS

Many factors influence the mathematical output; for example, the weighting factor and grid size used in calculations. It is also important to be aware of the factors influencing the image which is produced and its visual interpretation. Some of the features which should be considered are: grid size used for calculations, number of lines drawn on the surface, 'shape' of the diagram, i.e. height:width ratio. An example of a change in shape is given by comparing Figs 6 and 7 which have 1:1 and 1:2 ratio between *X* and *Y* axes respectively.

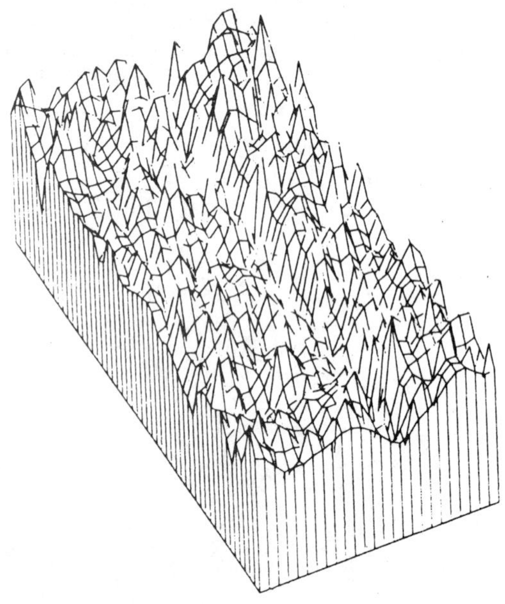

Fig. 7 — Three-dimensional plot of simulated surface in Fig. 6 with 1:2 ratio for *X*:*Y* axes.

For 3-dimensional plots, additional factors are scale on *Z* axis, elevation, azimuth. It should also be borne in mind when producing pictorial material of this type that it has been found that slopes are most easily identified if they have an angle between 30° and 60°, and those with a very steep or shallow angle are difficult to identify (see, for example, Cleveland & McGill 1987).

CONCLUSIONS

In this work, we have modelled data, produced representations in terms of 3-dimensional plots and contour maps which are as accurate as possible within the constraints of the package, simulated surfaces, and considered different sampling strategies.

Recently the goalposts have changed considerably as there have been some enormous advances in the instrumentation, and it is now possible to collect vast quantities of observations. The essential problem has therefore remained the same in that data need to be presented accurately. It has, however, changed drastically from one with very small numbers of observations to one with so much information that data reduction techniques will need to be considered. This is the area which is to be investigated next.

REFERENCES

Baird, E. M. (1988) The application of Raman spectroscopy to studies of elastohydrodynamic contacts. PhD thesis, Newcastle Polytechnic.

Cleveland, W. S. & McGill, R. (1987) Graphical perception: the visual decoding of quantitative information on graphical displays of data. *JRSS A* **150** 192–229.

16

Statistical analysis of a direction-dependent planar point process

A. B. Lawson
Department of Mathematical & Computer Sciences, Dundee Institute of Technology, Dundee Scotland

A number of examples of point process data which are related to fixed points, are examined. Two approaches to the analysis are reviewed: a likelihood approach and a MAP estimation method. The techniques can be extended to data which have been regionalized and consist of counts. Examples from epidemiology, ecology, and extensions to spatial quality control are reviewed.

INTRODUCTION

Point process data arise in a wide variety of applied research problems. Examples are found in reliability assessment, hydrological analysis, ecology, and epidemiology. Often the point process occurs in isolation, and the structure of the process itself is of prime concern (see, for example, Diggle 1983). However, in certain cases a process of points can be related to another process. Examples of this occur where there is a spatial or temporal relation between events of either process, e.g. crystal growth (point process) and lineament (line process) in geology (Berman 1986).

A special case of the above example is where a point process is related to a fixed point. Figs 1 and 2 show two examples of such point processes.

Example A (Fig. 1) represents the death certificate addresses of victims of respiratory cancer around a foundry in central Scotland (Lloyd 1982). The pattern of deaths was thought to be related to the central foundry stack location. Here the points could have arisen because of the effect of an environmental pollutant. Hence hypotheses concerning the detection of a spatial association with the central point are important.

Example B (Fig. 2) concerns the locations of fungi (Sporophores) found to be growing around a birch tree (Last *et al.* 1984). The sporophore pattern varies from year to year, but can be related in a relatively simple way to the central tree location.

Both the above examples are characterized by slowly varying patterns and non-

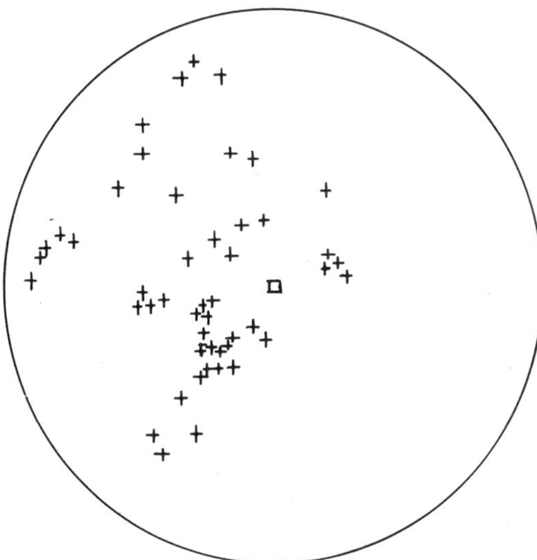

Fig. 1 — Death locations around smelter complex; Armadale, West Lothian. (Example A).

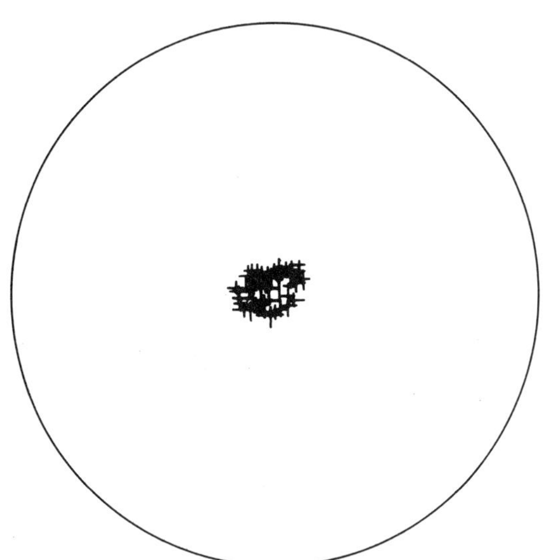

Fig. 2 — The pattern of Hebeloma sporophores in 1975 around a birch tree. (Example B).

stationarity. Hypotheses concerning these patterns will focus on describing this non-stationarity. Hence we will consider models for spatial trend, rather than second order structures.

STATISTICAL MODELS

To describe the spatial trend in these processes we can use a heterogeneous Poisson process model (HEPP). This model allows a flexible variety of slowly-varying forms for the intensity measure governing the point process.

We define:

$$\lambda(\mathbf{r}) = \lim_{|\mathbf{dr}|} \rightarrow 0 \left\{ \frac{E[N(\mathbf{dr})]}{\mathbf{dr}} \right\}$$

where $\mathbf{r} = (r, \theta)$ polar coordinate vector $\lambda(\mathbf{r})$ can take a variety of bivariate forms in r and θ. Commonly, we use:

$$\lambda(\mathbf{r}) = f(r) \,.\, g(r, \theta)/r \tag{1}$$

where f and g are suitable functions of r and θ. f and g can be *pdfs*, but this is not required, as a normalizing constant will be employed. We use a Weibull or gamma form for f and a von Mises form for $g(r, \theta)$.

We can estimate the parameters of $\lambda(\mathbf{r})$ by maximum likelihood via

$$l_1 = \sum_{i=1}^{n} \ln \lambda(\mathbf{r}i) - \Omega(A) \tag{2}$$

where the normalizing constant over the window of area A is

$$\Omega(A) = \int_{|A|} \lambda(\mathbf{r})\mathbf{dr} \tag{3}$$

STATISTICAL ANALYSIS OF A DIRECTION-DEPENDENT PLANAR POINT PROCESS

For a circular window we record all points (r_i, θ_i) and substitute into (2) for numerical maximization, the evaluation of (A) being a double integral. Note that the evaluation of (3) is simple because we are using a fixed central point.

For the more general problem of two processes this evaluation must be performed numerically.

Note also that if $\Omega(A)$ can be easily evaluated, a large variety of forms for $\lambda(\mathbf{r})$ are available. In fact, it has been found that models which describe clustered patterns can also be estimated via (2) where $\lambda(\mathbf{r})$ becomes a *conditional* intensity (Berman & Turner 1988, Penttinen 1984, Ripley 1988, p. 53).

For examples (A) and (B), we have used the models

$$f(r) = \lambda\delta r^{\delta-1} \exp[-\lambda r^{\delta}] \tag{4}$$

$$g(r,\theta) = \exp[(\kappa + \psi r)\cos(\theta - \mu_0)]/2\pi I_0(\kappa + \psi r) \tag{5}$$

This yields a five parameter model with a truncated Weibull distance decay and a von Mises density for θ. The ψ parameter allows a distance–angular correlation to occur. This could be appropriate when concentration at large distances are expected.

Figs 3 to 6 display various simulated realizations of the models, to demonstrate

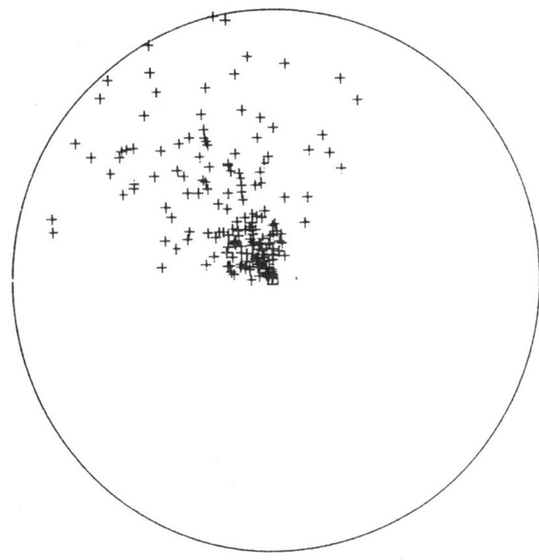

Fig. 3 — HEPP simulation of high angular concentration: $\lambda = 1$; $\delta = 1$; $\kappa = 10$; $\psi = 0$; $\mu_0 = 2.0$; $n = 200$.

their flexibility. Table 1 shows the maximum likelihood estimates for the interaction model applied to examples A and B. To demonstrate the form of these fitted models a single realization of the ML has been simulated, and these are shown in Figs 7 and 8.

The fit of the above models has also been reviewed by residual analysis. For spatial point process problems there is no unique definition of a residual, and we have

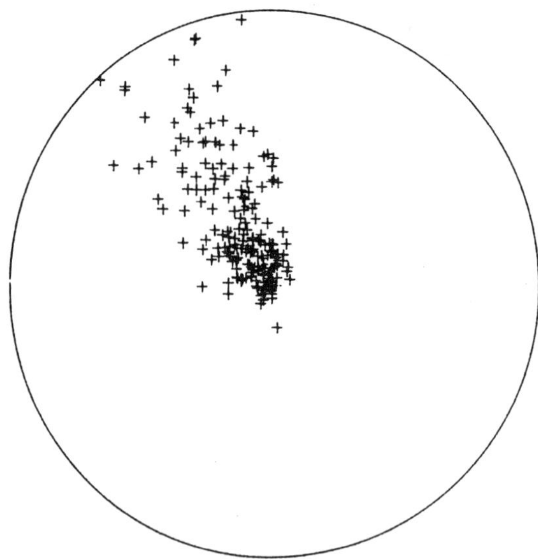

Fig. 4 — HEPP simulation of high interaction: $\lambda = 1$; $\delta = 1$; $\kappa = 0$; $\psi = 10$; $\mu_0 = 2.0$; $n = 200$.

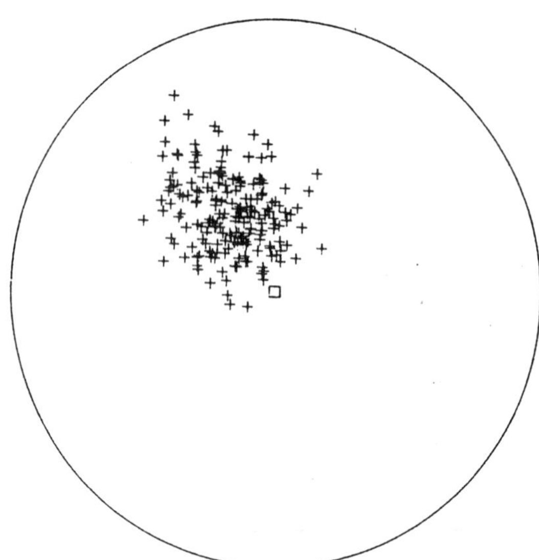

Fig. 5 — HEPP simulation of distance peaked-interaction: $\lambda = 1$; $\delta = 3$; $\kappa = 0$; $\psi = 10$; $\mu_0 = 2.0$; $n = 200$.

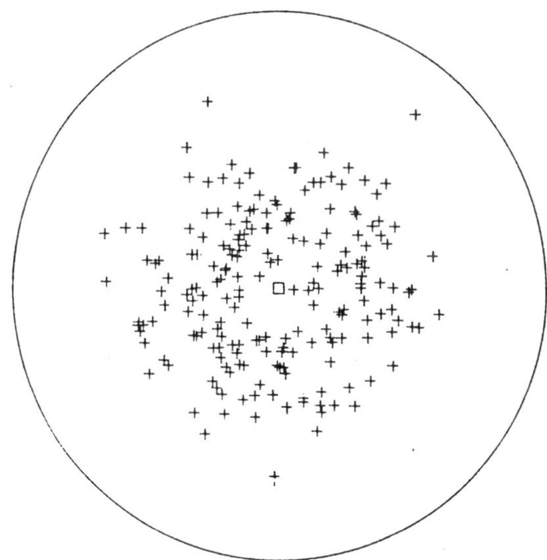

Fig. 6 — HEPP simulation of distance peak with uniform angular distribution: $\lambda = 1$; $\delta = 3$; $\kappa = 0$; $\psi = 0$; $\mu_0 = 2.0$; $n = 200$.

Table 1 — Likelihood results

Example A

λ	δ	κ	ψ	μ_0
0.556 (3.8×10^{-10})	2.23 (9.1×10^{-10})	0.0 (0.694)	1.306 (0.686)	3.024 (0.159)

Example B

λ	δ	κ	ψ	μ_0
186.2 (1.7×10^{-9})	3.85 (6.9×10^{-12})	0.0 (0.554)	4.486 (2.416)	2.625 (0.133)

(standard errors of estimates are given in brackets).

had recourse to an annular–sector grid and binning the points. We compared the observed counts to expected counts based on

$$\text{Expected count} = N \int |a_i| \frac{\lambda(\mathbf{r})\mathrm{d}\mathbf{r}}{\int |A|^{\lambda(\mathbf{r})\mathrm{d}\mathbf{r}}}$$

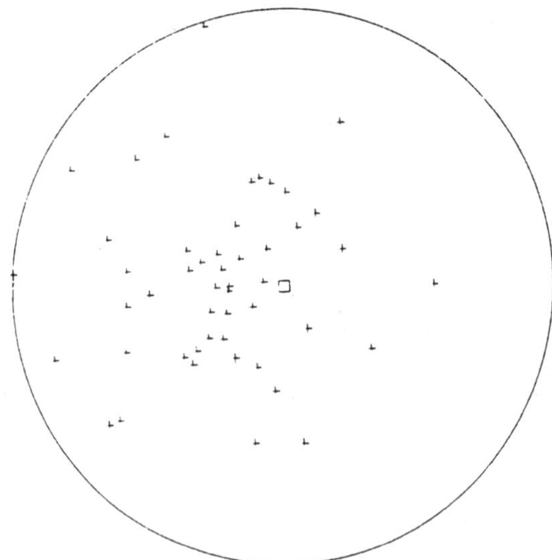

Fig. 7 — A single realization of the Armadale example, using the ML estimates for the interaction model (Table 1).

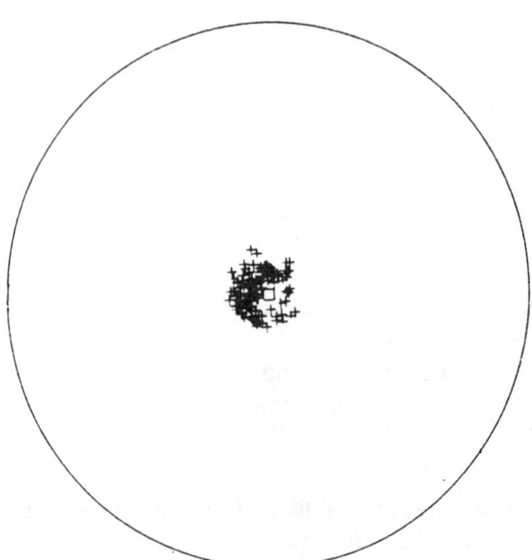

Fig. 8 — A single realization of the Hebeloma example, using the ML estimates for the interaction model (Table 1).

and used Anscombe residuals in visual inspection and normal plotting.

In Example B it was found that, occasionally, isolated clustered residuals could reduce a model fit. Hence, it appears that sporophores do not always behave smoothly. However, the 1975 example used here did not show such clustering. Example A does not display clustered residuals.

MORE COMPLEX MODELS

One criticism of the above models is that they allow only slowly-varying patterns to be modelled. This can be overcome by inclusion of covariate terms in $\lambda(\mathbf{r})$ which allow for environmental heterogeneity or for underlying periodicity in the point pattern. This latter effect could be due to an unknown environmental variation. The former effect could be a known variation. For the case of unknown variation two approaches could be used. First, harmonic terms could be added to $\lambda(\mathbf{r})$ to allow for regular variation. Second, a prior spatial distribution for the points could be assumed; this option is usually more favourable as it is parsimonious.

We can model spatial correlation in the point intensities by assuming that the intensity measure is a realization of a spatial stochastic process (Cox process). One method for doing this is to define

$$\lambda(\mathbf{r}) = \exp\{g(r)\}$$

where $g(r)$ is a spatial Gaussian process, hence $\lambda(\mathbf{r})$ a lognormal process. Any realization of $g(\mathbf{r})$ will be MVN. Hence, we can use a MVN prior for $g(\mathbf{r})$ and derive a posterior likelihood for the data. Using maximum *a posteriori* estimation (MAP) a two-stage estimation method is derived:

(1) estimate $g(\mathbf{r})$ from initial estimates of the vector of trend parameters $\boldsymbol{\beta}^0 = (\lambda, \delta, \kappa, \psi, \mu_0)$
(2) re-estimate $\boldsymbol{\beta}$ by GLS
(3) iterate between (1) and (2) until convergence.

This method can be implemented by using standard OLS methods for step (2). Figs 9–12 show examples of simulated realizations of the above process for a variety of prior distributions and trend vectors. The effect of the prior is to allow a background variation in the process, upon which the trend is superimposed. This is an important feature of the epidemiological example where *relative* risk is important.

A SPATIAL QUALITY CONTROL EXAMPLE

There are a number of possible applications for the above methods, not necessarily related to point processes. One such example could be in monitoring of material consistency where trends or clusters of impurities (points) must be detected in production sheets. Any process whose output is planar can be monitored for defects and can be checked for such spatial trends. Fig. 13 represents a diagonal section of a

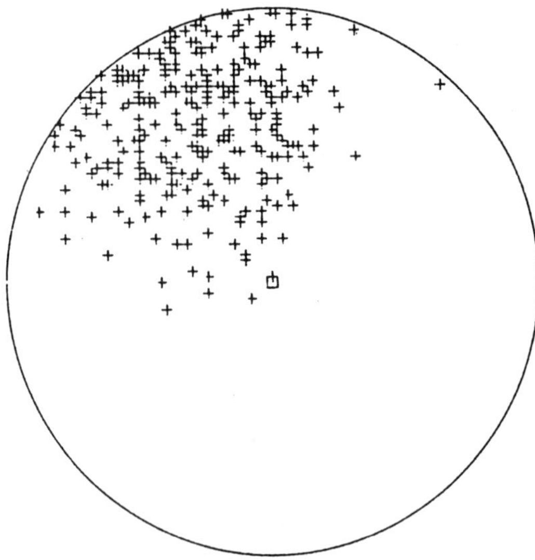

Fig. 9 — Thinned process with log-Gaussian intensity. Covariance range: 0.001; $\sigma^2 = 0.01$; $\lambda = 1$; $\delta = 1$; $\kappa = 0$; $\psi = 1$; $\mu_0 = 2$; $n = 300$.

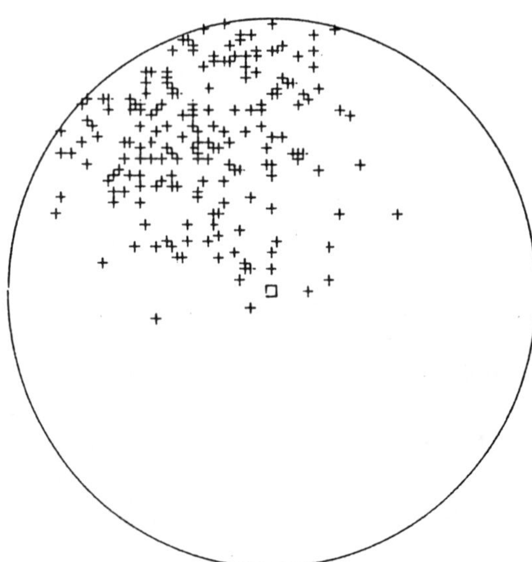

Fig. 10 — Thinned process with log-Gaussian intensity. Covariance range: 0.001; $\sigma^2 = 1.0$; $n = 300$. Other parameters are as above.

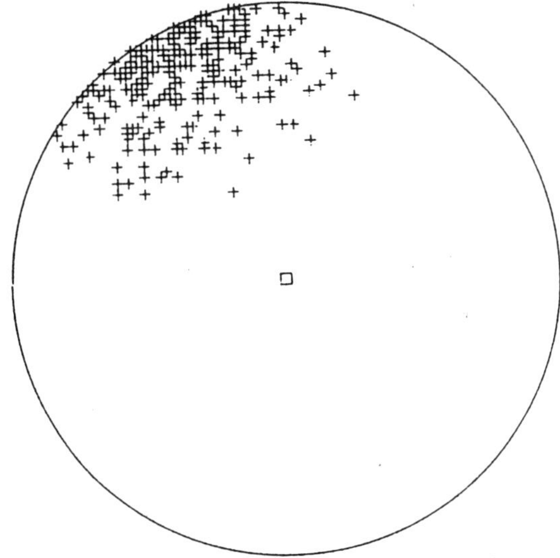

Fig. 11 — Thinned process with log-Gaussian intensity. Covariance range: 0.001; $\sigma^2 = 0.01$; $\lambda = 2$; $\delta = 5$; $\kappa = 0$; $\psi = 2$; $\mu_0 = 2$; $n = 300$.

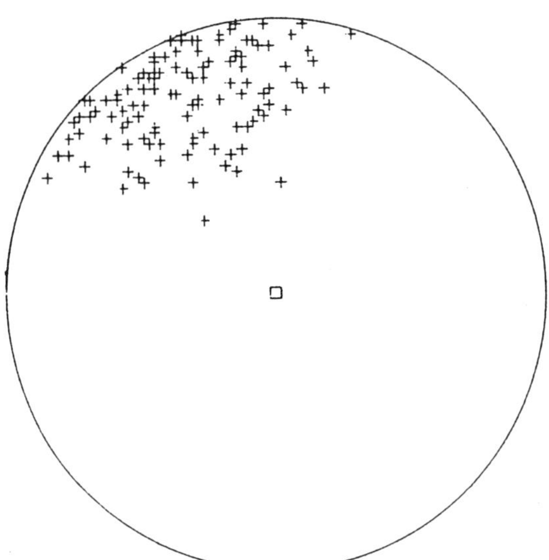

Fig. 12 — Thinned process with log-Gaussian intensity. Covariance range: 2; $\sigma^2 = 5$. Other parameters are as above.

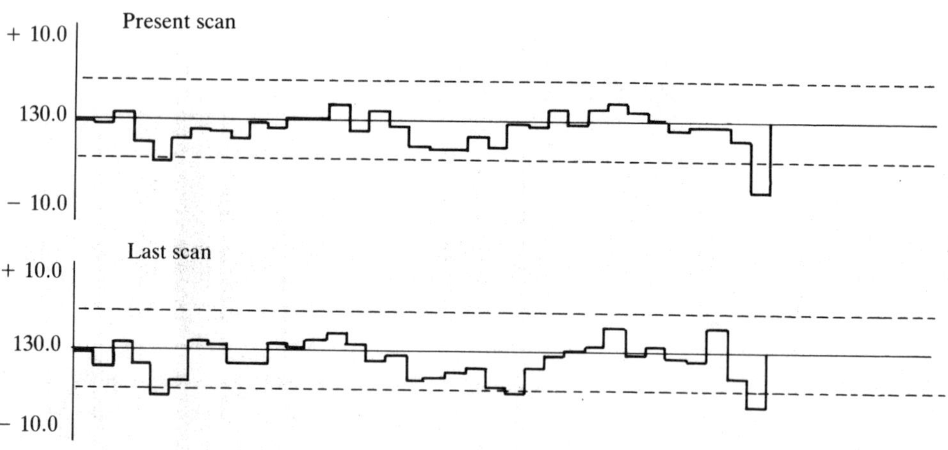

Fig. 13 — Scan output from a slit-film monitor measuring thickness of a synthetic sheet. Symmetric tolerance limits are shown.

slit-film technology extrusion process. The process extrudes sheets of synthetic material which are subsequently slit into fibre. Fig. 13 is part of a continuous thickness scan of a sheet. The thickness variation is averaged over short distances, and hence produces histogram-like peaks. The process contains spatial correlation and variance (possibly Gaussian?), and some trending (non-stationarity). If such a process is monitored for threshold crossings (high or low points), then these could be considered as a point process.

CONCLUSION

A number of models can be derived for assessing trend (non-stationarity) in a planar point process. The simpler HEPP models can be fitted, using standard likelihood methods. In fact, it is possible to use GLIM (Lawson 1988, 1989). The disadvantages of these models can be overcome by using a spatial prior for the linear predictor. With MAP estimation, a two-stage estimation procedure is possible. A variety of industrial quality control problems could be examined by using these methods, and it is hoped that this contribution will encourage their use.

REFERENCES

Berman, M. (1986) Testing for spatial association between a point process and another stochastic process. *Applied Statistics* **35** (1) 54–62.

Berman, M. & Turner, T. R. (1988) Approximating point process likelihoods using GLIM. (Submitted to *Applied Statistics*).

Diggle, P. J. (1983) *Statistical analysis of spatial point patterns*. Academic Press.

Last, F. T., Mason, P. A., Ingleby, K. & Fleming, L. V. (1984) Succession of fruitbodies of sheathing mycorrhizal fungi associated with *Betula pendula*. *Forestry Ecology and Management* **9** 229–234.

Lawson, A. B. (1988) Fitting the von Mises distribution using GLIM. *Jour. Applied Statistics* (Special issue: *Analysis of Circular Data*) **15** (2) 225–260.

Lawson, A. B. (1989) The von Mises distribution on GLIM. *Glim Newsletter* **19** December 1989.

Lloyd, O. (1982) Mortality in a small industrial town. In: Gardner, A. (ed.) *Current approaches to occupational health* **2** 283–309. Wright PSG.

Penttinen, A. (1984) Modelling interaction in spatial point patterns: parameter estimation by the maximum likelihood method. *Jyvaskyla studies in Computer Science, Economics and Statistics*, 7.

Ripley, B. D. (1988) *Statistical inference for spatial processes*. CUP.

Index